12⁵⁰/61

No. 1757
$24.95

THE COMPLETE BATTERY BOOK

RICHARD A. PEREZ

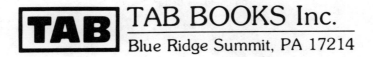

TAB BOOKS Inc.
Blue Ridge Summit, PA 17214

FIRST EDITION

FIRST PRINTING

Copyright © 1985 by Richard A. Perez

Printed in the United States of America

Reproduction or publication of the content in any manner, without express permission of the publisher, is prohibited. No liability is assumed with respect to the use of the information herein.

Library of Congress Cataloging in Publication Data

Perez, Richard.
The complete battery book.

Includes index.
1. Electric batteries. I. Title.
TK2896.P47   1985   621.31'242   85-9799

ISBN 0-8306-0757-9
ISBN 0-8306-1757-4 (pbk.)

Contents

Acknowledgments   v

Introduction   vi

1 What Is a Battery?   1
Definition of Terms—How Batteries Store and Transfer Energy—Battery Efficiency—Battery Types—How Cells Are Assembled into Batteries—Cell Polarity Nomenclature

2 Lead-Acid Batteries   14
Automotive Starting Batteries—Low Antimony "Deep Cycle" Batteries—True Deep Cycle High Antimony Batteries—Gel Cells—Lead-Acid Battery Characteristics—Charging the Lead-Acid Battery—Discharging the Lead-Acid Battery—Lead-Acid Battery Technical Data—Listing of Lead-Acid Deep Cycle Battery Manufacturers

3 Nickel-Cadmium Batteries   41
Sintered Plate Ni-Cads (Vented and Sealed)—Pocket Plate Ni-Cads (Vented and Sealed)—Ni-Cad Battery Characteristics—Charging the Ni-Cad Cell—Discharging the Ni-Cad Cell—Technical Data for Ni-Cad Cells—Ni-Cad Battery Manufacturers

4 Edison Cells   63
Physical Construction—Applications for Edison Cells—Edison Cell Cost—Life Expectancy—Edison Cell Characteristics—Charging the Edison Cell—Discharging the Edison Cell—Technical Data for the Edison Cell

5   Primary Cells — 70
Zinc-Carbon Cells—Alkaline-Manganese Cells—Mercury Cells—Silver Oxide Cells—Zinc-Air Cells—Lithium Cells—Listing of Primary Cell Manufacturers

6   Methods and Machines to Charge Batteries — 96
Lead-Acid Battery Chargers—Nickel-Cadmium Battery Chargers—120 VAC Powered Chargers—Solar Cells—A Motorized Charger for 12 Volt Systems—Wind and Water Power Sources—Small Ni-Cad Chargers

7   Using Batteries Effectively — 107
Wiring—Connections, Switches, Fuses, and Outlets—Internally Cross-Wiring the Battery Pack—How to Use Wire for Current Measurement—The Battery's Location in the System—Safety Requirements—Battery Instruments—Filtration

8   Inverters — 124
How the Inverter Works—Different Types of Inverters—Proper Inverter Sizing—Types of Batteries Suitable for Inverter Use—Inverter to Battery Interconnect—Inverter Location

9   Energy Management — 134
How to Estimate Energy Consumption—How to Size the Inverter—How to Size the Battery Pack—Energy Management Techniques—Appliances—Using Energy Management to Reduce System Cost

10   Developing Battery Technologies — 159
Nickel-Zinc Cells—Zinc-Chlorine Cells—Sodium-Sulphur Cells—Lithium-Metal Sulphide Cells—REDOX Cells—The Ideal Cell

Appendix A   Formulas and Conversion Factors — 175

Appendix B   Forms — 177

Glossary — 181

Index — 184

Acknowledgments

This book would not have been possible without the help of many people. Many users of battery stored energy have contributed information to this project. I wish to extend special thanks to: Stan Krute, Brian Green, Dale Hodges, Jim & Laura Flett, Dave Winslett, Dave Wilmeth, Scott & Stephanie Sayles, Larry Crothers, and Alan Trautman. Special credit goes to the Wizard for the wonderful charts and graphs. Thanks to Carolyn Moenning for proofreading the manuscript. I wish to personally acknowledge the support and encouragement of my loving wife, Karen, during this long project.

Many thanks to the battery manufacturers who have contributed information to this project. Among these are: Duracell Inc., C&D Batteries Inc., Trojan Batteries Inc., SAB NIFE Inc., Altus Corp., Panasonic Inc., Exide Corp., Eagle-Picher Inc., Varta Batteries Inc., Gates Energy Products Inc., Saft America Inc., and Power Sonic Corp. Special credit goes to the Heart Interface Corp. for their wonderful inverters which have done so much to widen the utility of battery stored power.

Introduction

Batteries are mysterious objects; their workings are on the sub-microscopic level of electrons and molecules. This book is an attempt to take the mystery out of the battery and reveal the nature of its inner workings and operating characteristics. The more we understand the processes of energy storage in batteries, the better we will be able to effectively use these batteries.

The information in this book will enable you to choose the proper battery for any particular job, to size this battery, and then to successfully apply this battery. All details of battery use are fully covered. This information is necessary for all users of battery stored electricity.

Batteries have been in use since the early 1800s. The electrochemical basis of the battery is the same now as then. Manufacturing technology and consumer demand have resulted in many types of cells, each with its own particular set of operating characteristics and suitable applications. This work discusses all commercially available types of batteries, both nonrechargeable and rechargeable. Details of the chemical workings and physical construction of each type are provided. The operating characteristics of each type are fully discussed and suitable applications are recommended.

Information regarding the proper cycling techniques for rechargeable batteries is covered in detail. The emphasis is on longevity and cost effectiveness. Proper use can greatly extend the lifetime of expensive rechargeable cells. All the different types of batteries are rated on the basis of cost.

Users of alternative energy systems will find a wealth of material on battery application and cycling. The process of energy management is discussed in great detail. Energy management assures that alternative energy people get the most bang for the buck. Techniques for sizing the battery and all other system components are discussed.

This book is a complete survey of batteries and their application. It assumes only a minimal knowledge of science. This information has not been available before in a single volume.

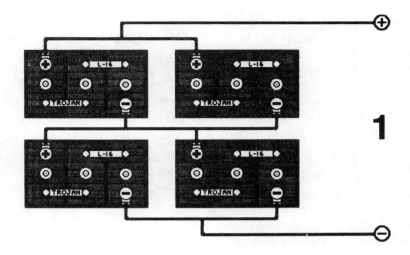

What Is a Battery?

A battery is a device that stores electrical energy. Batteries are chemical machines. In the battery, chemical energy is converted into electrical energy. Electricity is stored within the battery in the form of potential chemical bonding between the battery's active materials. As a battery is charged or discharged its chemical composition changes.

In some batteries the chemical reaction is not reversible. This type may only be discharged. It cannot be recharged. Batteries which cannot be recharged are known as *primary* batteries. One example of a primary battery is the disposable zinc-carbon cell used in flashlights.

Other types of batteries are rechargeable. The chemical reaction within a rechargeable battery is reversible. Rechargeable batteries are known as *secondary* batteries They may be emptied and refilled many times. An example of a secondary battery is the lead-acid battery used to start an automobile.

Batteries are chemical engines used to push electrons around. The electron is a very small negatively charged particle which revolves around the center of all common atoms. Electricity is electrons in motion.

DEFINITION OF TERMS

In order to discuss batteries we need to understand some basic terms dealing with electricity. These terms are defined here so as to apply particularly to the type of electricity stored in batteries.

Voltage

The volt is the unit of measurement of electrical potential difference between any two points. This electrical potential difference is known as the *electromotive force*, or EMF. The volt is the unit of the electrical force.

Electronics will flow through a wire only if forced to do so. If one end of a wire has more electrons than the other end, there is an electrical potential difference between the two ends of the wire. Voltage is the electronic inequality between any two points. The greater this difference in the number of electrons from point to point on the wire, the greater the potential difference and the greater the voltage. The concept of voltage is roughly equivalent to the hydraulic concept of pressure.

Let's say we've got a garden hose and are watering the turnips. The water is flowing from the hose at a certain pressure. The higher the pressure the more water comes out of the hose per second. The lower the pressure the less water will run from the hose. The pressure in a hydraulic system is analogous to voltage in an electrical system. Voltage is the unit of electronic pressure. The physicist would call it the Electromotive Force or EMF.

The electrical system in automobiles uses a voltage of around 12 volts. The electrical systems in American houses use 120 volts while in Europe the standard voltage for houses is 240 volts. Long distance electrical power lines have voltages over 50,000 volts. Different systems use different voltages to accomplish different tasks.

Throughout this book we will primarily be concerned with low voltages. Most types of batteries have individual cell potentials of less than three volts. For example, a fully charged common flashlight battery (zinc-carbon) has a voltage of 1.5 volts. Batteries used in alternative energy applications are collections of cells having a total voltage of between 12 and 48 volts.

Current (Amperage)

The ampere is the unit of electrical current. It deals with how many electrons pass any particular point per unit time. The ampere is the unit of the measurement of rate of electronic flow. The direction of the flow of electrons may be constant or it may change direction. Current that flows in a single direction is known as *direct current* or dc. A current which changes direction is known as *alternating current* or ac. Batteries store and transfer only direct current (dc).

Again the analogy with hydraulics applies to current. If voltage is equivalent to pressure then amperage is the same as rate of flow (in gallons per minute or whatever units of measurement). If the pressure (voltage) is turned up on the garden hose then the rate of flow (amperage) increases.

A standard scientific definition of the volt and the ampere is this: a voltage of one volt will cause one ampere of current to flow through a resistance of one ohm.

Resistance

Resistance is the property of materials to resist the flow of electrical current. All materials have resistance. Materials like glass, rubber, and plastic have very high resistances and are known as *insulators*. Materials such as copper, aluminum, and most metals have very low resistance and are known as *conductors*.

The idea of resistance compares to the idea of hose diameter in the hydraulic analogy. The larger the diameter of the hose, the less resistance is offered to the stream of water. This results in a greater flow of water.

In electricity, the unit of resistance is the ohm (Ω). One ampere of current will flow through a one-ohm resistance when a voltage of one volt is applied to the resistor.

The relationship between voltage, current and resistance is best expressed in the equation known as Ohm's Law.

$$E = IR$$

E = voltage expressed in volts
I = current expressed in amperes
R = resistance expressed in ohms

This relationship is used constantly throughout physics, electronics, and this book. An understanding of the principles expressed in Ohm's Law is critical for working with electricity.

Power

Power is the product of voltage and current. The unit of power is the watt. One watt of power is consumed when 1 ampere of current flows at one volt.

$$P = IE$$

p = power expressed in watts
I = current expressed in amperes
E = voltage expressed in volts

The watt is the measurement of how much power is being transferred. For example, if we replace a 40-watt lightbulb with an 80-watt light bulb we will use twice as much power. Since the voltage in our lamp is constant, the number of amperes of current through our bulb will double.

Power is dependent on both amperage and voltage. Consider a light bulb rated to consume 120 watts of power. If the bulb is powered with 120 volts (standard house voltage), it will use 1 ampere of current. If the same wattage bulb is run on 12 volts (as in a car), it will use 10 amperes of current to consume 120 watts.

Most battery systems run on low voltages (48 volts or less). The amount of current needed to produce a given amount of power is larger than in regular household circuits (120 volts). This factor is important since losses in power due to resistance in wiring and switches is primarily a function of the amount of current flowing. In low voltage systems we need big wire and low loss switching.

Capacity

Capacity is how much electrical energy the battery will contain. The unit of capacity is the ampere-hour. The larger the ampere-hour rating of the battery the larger its capacity. The ampere-hour is the product of the amount of current a battery will deliver and the time over which it will deliver this current.

A battery with a capacity of 100 ampere-hours will deliver 1 ampere for 100 hours. The same battery will deliver 10 amperes for 10 hours, or 100 amperes for 1 hour. Most commercial lead-acid batteries are rated at a 20-hour discharge rate and at 78° F.

Automobile batteries have capacities between 60 to 100 ampere-hours. Large storage batteries in alternative energy systems have many thousands of ampere-hours. Flashlight batteries vary in capacity from 0.5 ampere-hours to 10 ampere-hours. The physical size and weight of a battery is roughly equivalent to its capacity.

State of Charge

The state of charge of a battery tells how much of the battery's capacity is available. A battery which has its entire capacity available is said to be at a 100 percent state of charge. A battery which has had half its capacity removed is said to be at a 50 percent state of charge. A battery which has had its entire capacity withdrawn is at 0 percent state of charge. The state of charge of a battery is important because it tells us when it is empty and needs refilling. It also tells us when it is full and when to stop charging.

The charge condition of the battery is sometimes expressed as *state of discharge*. The state of discharge indicates how much energy has been withdrawn from the battery. State of charge indicates how much energy is remaining in the battery. A battery at a 80 percent state of charge will have had 20 percent of its energy withdrawn. It would have a state of discharge of 20 percent.

Efficiency

Efficiency is the ratio of how much energy we get out of a system to how much energy we put in. Efficiency = Power Out / Power In. No known system is 100 percent efficient. Energy is lost in all conversion, transfer, and storage processes and batteries are not exceptions to this rule. It takes more energy to charge a battery than we can get out of the battery. Batteries vary in efficiency depending on the chemical processes used, the particular battery type, and the conditions of service. Efficiencies of the various types of batteries are discussed under each type.

Specific Gravity

The specific gravity of a material is its density divided by the density of water. Density is the ratio of the mass of a material to its volume. A material which has a density two times that of water has a specific gravity of two.

In lead-acid batteries the specific gravity of the electrolyte is of importance because it is an indicator of the state of charge. In other types of batteries (nickel-cadmium) the electrolyte specific gravity does not relate to state of charge. The *hydrometer* is the tool used to measure the specific gravity of the electrolyte.

Rate of Charge or Discharge

The rate of charge or discharge of a battery is expressed in terms of the battery's capacity. This is done even though the rate of charge or discharge is a current which is actually measured in amperes. This is important and confusing. The charge or discharge rate is expressed in amperes as the battery's rated capacity divided by a time factor. This time factor is the amount of time during which the battery is cycled. As an equation it looks like this:

$$I = \frac{C}{T}$$

I = Rate of charge or discharge expressed in amperes
C = Battery's rated capacity expressed in ampere hours
T = Cycle time period expressed in hours

For example, consider a fully charged battery with a capacity of 100 ampere-hours. If this battery is totally discharged within a ten-hour period, then the rate of discharge is ten amperes. Such a rate of discharge is known as a C/10 rate. If the same battery is discharged within a 50 hour period, then the rate of discharge is two amperes, or C/50. The same format refers to the charge portion of the cycle. A battery which was fully discharged and is filled during a period of ten hours is being charged at a C/10 rate.

Rates of charge and discharge in batteries are commonly referred to as ratios between capacity and time. The actual amount of current used in each particular case is dependent on the battery's capacity. This allows us to express rates of charge and discharge in general terms rather than as specific quantities of current.

HOW BATTERIES STORE AND TRANSFER ENERGY

The battery converts chemical energy into electrical energy. In rechargeable batteries the conversion process is reversible. Rechargeable batteries can also convert electrical energy into chemical energy.

The Cell

The conversion and storage processes take place in the basic building block of all batteries—the cell. The cell contains the active materials and the electrolyte of the battery. Most batteries are composed of many cells because the voltage potential of each chemical cell is quite low (a few volts at most). The electrical storage capacity of a cell is roughly proportional to its physical size. The larger the cell the more capacity it has. A battery is composed of cells which are assembled together to increase the voltage or the capacity of the battery.

Active Materials

The cell contains two active materials which can react chemically to release free electrons (electrical energy). Such materials are known as *electrochemical couples*. The active materials are usually solid. The cell also contains an electrolyte which aids in the transfer of the electrons between the electrochemical couples. The electrolyte is usually a liquid, jelly, or a paste. In some cells, the electrolyte participates in the chemical reaction in addition to acting as a path for electrons.

During the discharge of a cell, the active materials undergo chemical reactions which release free electrons. During this reaction the chemical compositions of the active materials are changed. The reactants actually become different chemical compounds. When all the original active materials have undergone reaction, the cell will produce no more free electrons. The cell is "dead."

In the rechargeable secondary cell the chemical process is reversible. By forcing electrons through the cell in the opposite direction, the active materials can be restored to their original chemical composition. This is known as "recharging" the cell.

The cell has polarity: one of the active materials is electron deficient and is positive. The other active material is electron rich and is negative. The flow of electrons from the cell is from the negative pole (cathode) to the positive pole (anode). During recharging the flow is reversed—the electrons flow from the anode to the cathode.

There are many different chemical compounds which form electrochemical couples. The electrical nature of the cell is determined by the electrochemical couple used. Due to restrictions such as material cost, technical limitations, and material availability, relatively few electrochemical couples are actually used in commercially available batteries. Two examples of electrochemical couples commercially produced are the lead-acid reaction and the nickel-cadmium reaction. Details of these reactions are available in Chapters 2 and 3.

Energy Storage in Chemical Reactions

A charged battery has energy stored within its chemical bonds. The active materials (the electrochemical couple) within the charged battery exist in such a form that the reaction between the materials releases free electrons. These free electrons are available for our use at the battery's output terminals.

All elements have electrons revolving around a nucleus of protons and neutrons. Chemical bonding between elements is the exchange or sharing of these electrons. For example, sodium and chlorine are chemical elements. They are distinct materials, each with its own distinct characteristics. When they bond with each other they become salt, which is another totally distinct material. Here is a case of two elements (sodium and chlorine) chemically bonding to form a compound (salt).

When this bonding occurs the sodium atom gives up an electron to the chlorine atom. Each atom becomes electrically unstable; they become ions. These ions cling to each other from electrostatic attraction and the resulting compound is more stable than the original elements it is made up of. The entire two-atom system has less energy. Atoms form ionic chemical bonds in order to reach states of greater electrical stability.

The science of chemistry deals with the nature of the elements and the myriad forms of bonding which can occur between them. In all chemical reactions which release energy, the materials bond in order to form a more stable structure. The idea is similar to the fact that water runs downhill. It seems that all the materials around us are seeking to form structures of the lowest energy potential—to become more stable. In batteries the active materials can form more stable structures of lower energy by transferring electrons. The active materials in batteries may be either elements or compounds. In some cases the electrochemical couples are both. In all batteries all the energy stored within them is not available to the user. There are many areas of loss, the greatest being heat.

Discharging

The addition of a load to the battery's output terminals allows the electrons to be transferred between the active materials. This process is known as discharging. The electrons flow as the materials seek a more stable electrical configuration. The chemical nature of the active materials changes to one of a lower energy level.

All batteries tend to discharge themselves over a period of time. The electrochemical discharge reaction takes place in the absence of an external load to the battery. The path of the electrons during self-discharging is through the electrolyte. The path through the electrolyte is of much higher resistance than the path through an external load. If it were not, then we would not be able to use the energy from the battery; it would discharge itself faster than we could use the energy. This *internal resistance* of the battery is the major point of energy lost to heat in all modes of battery operation.

Charging

The charging process is simply the reverse of discharging. A voltage is applied across the battery's terminals causing electrons to flow through the battery. In order to overcome the battery's internal resistance the charge voltage must be higher than the output voltage of the battery. The direction of the electron flow is the reverse of that during the discharge cycle.

The reversal of this electron flow supplies the energy necessary to return the active materials to their charged state. The chemical bonds made during discharge are broken by the charging process. The active materials regain their higher energy state. They become the original chemical compounds found in a charged battery. The electrical energy is converted into chemical energy. Once again there is energy lost to heat in the conversion process.

BATTERY EFFICIENCY

The efficiency of a battery is the ratio of how much energy we can get out of a battery versus how much we put in. Efficiency = Power Out / Power In. No battery is 100 percent efficient. Most batteries fall in the 60 percent to 80 percent efficiency range. It is necessary to put about 30 percent more energy into the battery than we can get out of it. There is energy lost in storage, charging, and discharging. There are many factors affecting the efficiency of a battery, some are user-controllable and some are not.

The energy loss to heat, due to the internal resistance of the battery, is unavoidable. This is an inherent loss present in all types of batteries. The losses due to self-discharge can be controlled by proper battery cycling. All types of batteries have a finite lifetime. At the end of this lifetime they become more and more inefficient. Periodic replacement of the batteries is necessary to avoid low system efficiency.

Temperature

Temperature is an important factor affecting battery efficiency and performance. The rate of the chemical reaction in most batteries is highly temperature dependent. If the battery is too warm it will self-discharge very rapidly. If the battery is too cold it will not readily deliver its energy. Batteries are chemical engines, and as such, they are limited by the characteristics of their chemistry. Most electrochemical couples will only bond readily within certain temperature ranges.

Lead-acid type batteries are notorious for being sluggish in starting cars on cold mornings. Other types of batteries, nickel-cadmium for example, have wider operating temperatures. We can maximize the efficiency of a battery system by attempting to keep it at optimum operating temperatures. Details on the temperature preferences of each type of battery are covered in the section on that particular type.

Purity

Efficiency is also affected by the purity of the reactants in the battery's cells. Batteries that are manufactured from impure components will be less efficient and shorter-lived. Rechargeable wet cell batteries that have had their cells contaminated will not deliver energy efficiently or for long. Contaminants in the battery's cells chemically tie up the active materials and prevent them from participating in their intended reactions. This is critical to users of lead-acid batteries who open the cells to check the specific gravity of the electrolyte. Batteries are chemical engines—purity of the reactants is essential for efficient, long-lived service.

Sizing

Other major factors affecting efficiency are controllable by the user. If a battery is to perform properly it must have the proper amount of capacity to meet the need. Battery packs which are sized either too large or too small will be inefficient and not cost effective. The sizing of battery packs is discussed in detail in Chapter 9.

The sizing of the battery's capacity in relation to the demand for energy determines the rate of discharge. Smaller capacity batteries will have to deliver energy faster in relation to their size than will larger batteries. Proper sizing of the battery pack's capacity results in acceptable discharge rates in relation to the battery pack's capacity.

Rate of Energy Transfer

All batteries have limitations on the rate at which they may be cycled. If energy is transferred through the battery too rapidly, then the losses due to heat greatly increase. This additional energy lost to heat is not usable and places additional mechanical wear on the battery's physical structure. The grids within the cells expand and contract more with rapid cycling and they wear out more quickly. Some types of batteries will lose electrolyte to gassing if cycled to rapidly.

If a battery is to be efficient, if must not be either charged or discharged too rapidly. Different types of batteries have different limitations as to cycling rates. This information is given under each specific type. One factor that is common to all types is the capacity of the battery in relation to the discharge rate. A correctly sized battery will eliminate discharge rates which are too high.

BATTERY TYPES

Of the hundreds of electrochemical couples available for use in batteries, only a few are commercially available. The following information is intended to be a simple user's overview of the different characteristics of seven commonly produced secondary types. Table 1-1 provides the information in trade form. This information is highly generalized. Actual battery performance and specifications vary widely between manufacturers and different users.

The capacities in Table 1-1 are given in ampere-hours per cell. The figures for Energy Efficiency are rough averages. The efficiency of a battery depends in a large part on the user, since proper cycling techniques and sizing are required to maximize efficiency. All voltages given on the table are at 78° F. Cell voltages will vary with temperature.

The number of times a battery can by cycled and its calendar life vary widely. If they are cycled at the correct rates for each type then longevity is increased. If they are kept at the proper temperatures then longevity is increased.

The cost figures in Table 1-1 are given in dollars per kilowatt-hour ($/kWh), dollars per kilowatt-hour per cycle ($/kWh/cycle), and dollars per kilowatt-hour per year ($/kWh/year). These figures are approximate. Battery costs differ from manufacturer to manufacturer. Batteries which are not cycled properly will be more expensive to use.

Lead-Acid

The lead-acid couple is extensively manufactured in many forms. After 80 years of commercial production, the lead-acid couple is the least costly per kilowatt-hour. It is the most cost effective battery for motor starting and home energy storage. High antimony deep cycle batteries cost less per kilowatt-hour per year to use in deep cycle service than any other type.

Nickel-Cadmium

Nickel-cadmium batteries are manufactured in many sizes. The sealed sintered plate ni-cads have very small capacities. The cost figures on the table reflect this. The larger "wet" nicad is useful in home energy storage. It is approximately twice as expensive as a lead-acid system. This greater expense is justified in some cases by temperature and longevity considerations.

HOW CELLS ARE ASSEMBLED INTO BATTERIES

Most batteries we encounter are composed of more than one cell. In fact, the word battery means any set of devices arranged or used together. The term "flashlight battery" is incorrect when referring to a single flashlight cell. The cell is the basic indivisable unit. A battery is a group of cells.

Cells are combined in two configurations to increase the power of the battery. The first method of wiring the cells is in *series*. A series electrical circuit has only one path available for the electrons. In the series configuration each cell has its positive terminal attached to the negative terminal of another cell.

The second configuration is known as *parallel* wiring. In a parallel electrical circuit there is more

Table 1-1. Comparison of Secondary Battery Types.

Battery Family	LEAD-ACID				NICKEL-CADMIUM		
Type of Battery	Auto Starting	Diesel Starting	Deep Cycle	GEL Cell	Sealed Sintered	Sealed Pocket	Vented Pocket
Capacity Ah/Cell	33 to 340	60 to 570	180 to 2200	1 to 40	0.1 to 20	0.8 to 30	10 to 1400
Operating Voltage	2 VOLTS				1.25 VOLTS		
FullCharge Voltage	2.55 VOLTS				1.45 V.	1.60 VOLTS	
Cutoff Voltage	1.75 VOLTS	1.9 V.		1.8 V.	1.00 VOLT		
Capacity 104° F.	105 %			108 %	98 %		
Capacity 32° F.	70 %			87 %	90 %		
Capacity -20° F.	20 %			40 %	65 %		
Energy Efficiency	75 %				65 %		
Loss per Month.	New: 6 % Old: 50 %				10 %	5 %	
Cycle Life	150 to 250	500	1500 to 2000	500 to 1000	500	500 to 1000	1500 to 2000
Calendar Life-Float	2 to 5 years	8 years	5 to 15 years	2 to 5 years	10 years	15 years	20 years
Cost in $/kWh	70.	140.	95.	555.	900. to 10,000.	900. to 3600.	400. to 1200.
Cost in $/kWh/cycle	0.35	0.30	0.06	0.74	2. to 20.	1.20 to 4.80	0.22 to 0.70
Cost in $/kWh/year	20.	17.50	9.50	159.	90. to 1000.	60. to 288.	20. to 60.

than one path for the electrons to travel. In parallel configuration, the cells have their positive terminals interconnected and their negative terminals interconnected.

In Series for Voltage Increase

All chemical battery cells have low voltage outputs. The lead-acid cell has an output of about 2 volts. The nickel-cadmium cell has an output of 1.25 volts, while the zinc-carbon flashlight cell has an output voltage of about 1.5 volts. These are absolute limits on cell voltage. These limits are determined by the potential energy of the electrochemical reaction involved. Size is not a factor in the cell's output voltage. Making the cell larger simply increases its capacity, while the output voltage remains constant.

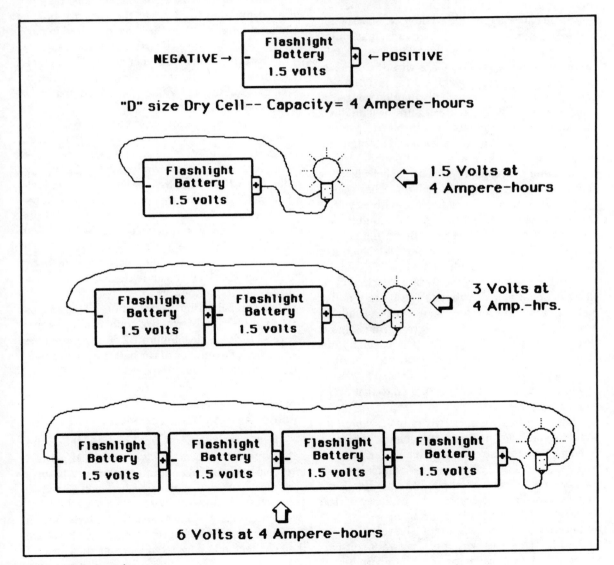

Fig. 1-1. Cells in series.

Electrochemical cells are interconnected to each other in series in order to use their stored energy at higher voltages. A group of interconnected cells is called a battery. If two cells are wired in series the resultant battery will have twice the voltage. If six cells are wired in series the resultant battery will have six times the voltage of a single cell. For example, an automotive starting battery consists of six lead-acid cells (each 2 volts) in series to give a resultant battery of 12 volts.

Some batteries contain all their cells in a single battery casing, some do not. Due to weight limitations, very large storage batteries are usually cased as single cells. These are wired in series to produce the appropriate voltage. In some large storage batteries, up to three cells may be housed in the same case. Larger batteries are broken down into smaller units for ease of transport and handling. The basic cell in large lead-acid storage batteries weigh between 40 and 400 pounds.

Another example of series use of cells is in the common flashlight. Two flashlight cells, each a zinc-carbon cell at 1.5 volts, are used in series to provide 3 volts to the bulb. If your flashlight takes four dry cells in series then the operating voltage of the bulb is about 6 volts. Figure 1-1 illustrates the series use of flashlight batteries.

A battery consisting totally of cells wired in series has one major drawback. The battery is like a chain; it is only as strong as its weakest link. In a series-wired battery the electrons must move through each and every cell. If one cell is discharged then the battery is inoperative, regardless of the condition of the rest of the cells. The output power of the entire battery is limited to that of the weakest cell.

Let's say that we have two batteries which we wish to combine in series for voltage increase. Assume that they are both 6-volt batteries (each with three lead-acid cells in series) which we wish to combine to get an output of 12 volts. Let's assume that one battery has the capacity of 100 ampere-hours and the other has a capacity of 300 ampere-hours. The resultant 12-volt battery formed by the series wiring of the two 6-volt batteries will have a capacity of 100 ampere-hours.

smallest cell within a series-wired battery pack determines the capacity of the pack. When the smallest cell is fully discharged it will not conduct any more electrons. In this state the series circuit is broken. The entire battery is dead, regardless of the state of charge of the rest of the cells.

Cells in Parallel for Capacity Increase

Cells or batteries (collections of cells) may be wired in parallel to increase the capacity of the resultant battery. When the cells are wired in parallel the voltage stays the same, but the capacity of the battery so formed is increased. The capacity of the resultant battery pack is the sum of the capacities of the individual paralleled batteries which make it up.

For example, assume that we have two 12-volt automotive batteries we wish to parallel to increase the capacity of the resultant battery pack (remember the voltage will stay the same—12 volts). Each 12-volt car battery is cased individually. In each case there are six lead-acid cells in series to produce the output voltage of 12 volts. Lets assume one 12-volt battery has a capacity of 100 ampere-hours and the other has a capacity of 60 ampere-hours. The resultant battery formed by paralleling the 12-volt car batteries will have a capacity of 160 ampere-hours.

In a parallel wiring configuration, all the anodes of the paralleled batteries are connected together, as are all the cathodes. Figure 1-2 demonstrates the paralleling of a number of car batteries to produce battery packs of larger capacity.

Series and Parallel Interconnection Used Together

In alternative energy applications, the entire battery pack may contain both series and parallel interconnection. Since alternative energy battery systems usually run on voltages between 12 and 48 volts, there is always series interconnection between batteries. In some cases, the batteries which have been used in series (for voltage increase) are then connected in parallel to increase the capacity of the entire battery system.

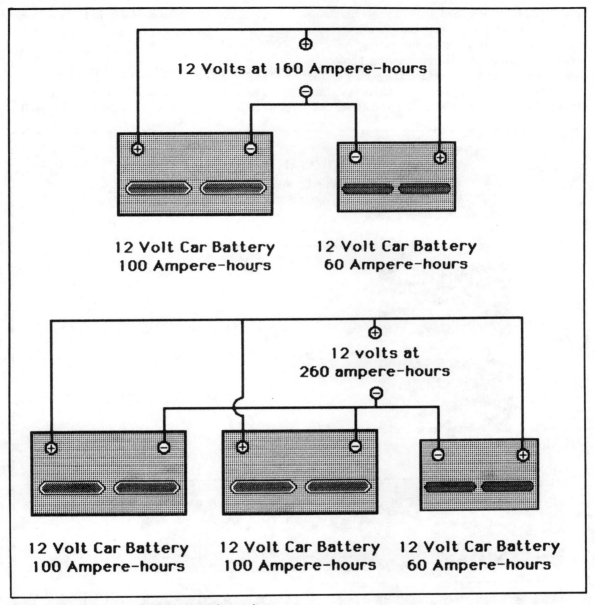

Fig. 1-2. Batteries in parallel for increased capacity.

Figures 1-3 and 1-4 illustrate some configurations used in home energy systems. The basic battery used as a building block in the illustration is the Trojan L-16. This is a 6-volt, 350 ampere-hour, high antimony deep cycle lead-acid battery. Each L-16 has three lead-acid cells in series, all enclosed within a single battery case. Each individual cell has a capacity of 350 ampere-hours. The L-16 is very popular with users of alternative forms of energy because of its cost effectiveness.

These same wiring techniques can be used to assemble battery packs of any desired voltage and

11

capacity. The wiring method is the same no matter what type of battery is in use. It is important to maintain a balance within the battery pack. It is highly desirable that all the individual batteries making up a pack be the same. They must be the same size, the same type, and the same age. Battery packs assembled from dissimilar batteries will not be efficient and are difficult to effectively charge.

CELL POLARITY NOMENCLATURE

There is some confusion as to the exact definitions of anode and cathode. These terms are de-

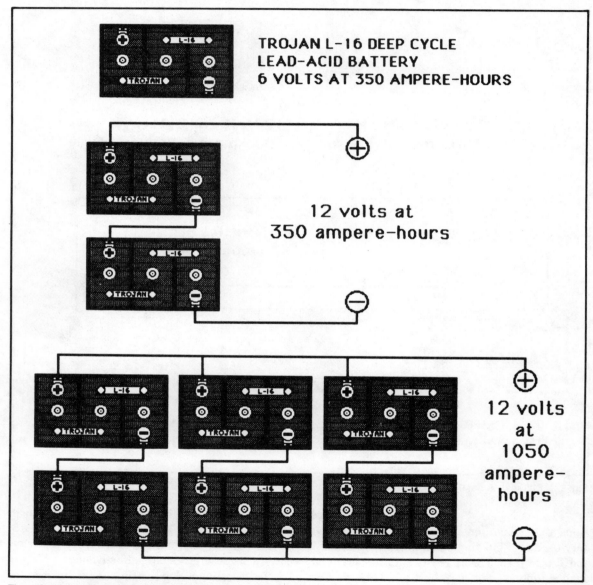

Fig. 1-3. Alternative energy battery packs for 12-volt systems.

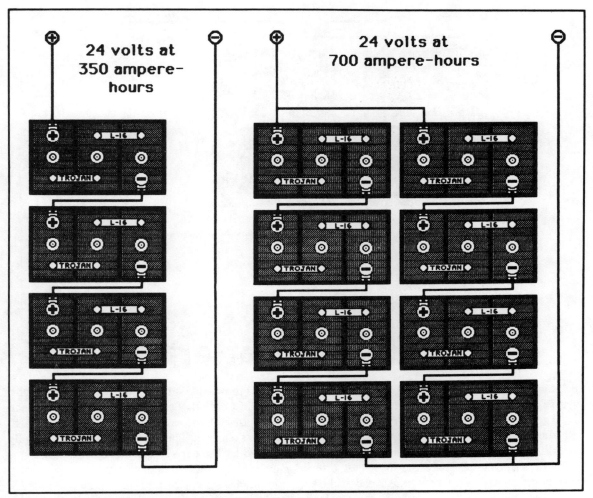

Fig. 1-4. Alternative energy battery packs for 24-volt systems.

fined differently by electricians and chemists. The electrician always considers the anode of a battery to be the positive terminal. The chemist, however, may sometimes consider the negative electrical terminal of the cell to be the anode. This difference of definition is due to the fact that the chemist always considers the oxidized material as the anode. During discharging, cells have their negative electrical terminal oxidized and as such, it is considered by the chemist to be the anode.

Throughout this book, the positive terminal of secondary cells is referred to as the anode. This is chemically true during the charging portion of the secondary cell's cycle. This convention is used with all material on secondary cells in this book.

All primary cells are referred to in the strict chemical definition of anode. The chemical anode of the primary cell is the negative electrical output terminal. This convention is used with all information referring to primary cells.

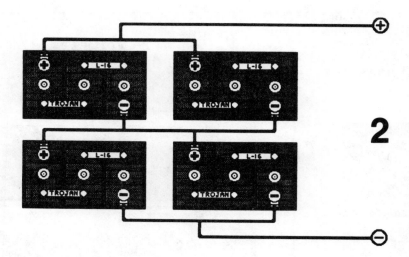

Lead-Acid Batteries

Lead-acid batteries have been in general usage for over 75 years and are the most common type of rechargeable battery. You are probably familiar with the one under the hood your car. Lead-acid batteries provide starting power for almost all of the world's vehicles and are the most cost effective type for home energy storage. All lead-acid batteries are based on the same sulphuric acid and lead chemical reaction. However, these batteries have evolved into different types for specialized needs. The capabilities of these different types very widely and need to be considered separately.

AUTOMOTIVE STARTING BATTERIES

The lead-acid automotive starting battery is a highly developed and specialized type of battery. It has evolved over the last 70 years to do only one job—start your car. It is not designed for deep cycling and is quickly ruined by being fully discharged repeatedly. The reasons for this lie in the battery's chemical and mechanical construction.

The lead-acid starting battery has been optimized to provide a large amount of energy for a very short period of time. The average car uses in excess of 300 amperes for just a few seconds to start the engine. The battery is then continually charged by the car's alternator until the motor is shut off. In order to deliver starting service at a minimum weight, size, and cost, these batteries are constructed of a large number of thin plates of lead sponge. The large number of thin sponge plates provide the maximum surface for chemical reaction. This large area gives the battery the ability to deliver high amperage while maintaining acceptable voltage levels. These batteries are designed for very shallow cycling. Under starting service this battery is normally cycled less than 1 percent of its rated capacity. Anyone who has had starting problems with an automobile and who has deeply and repeatedly cycled the starting battery is familiar with its rapid degeneration in deep cycle service.

The thin sponge plate construction results in very poor deep cycle performance. The starting battery begins failing after less than 100 cycles of

50 percent or more of its capacity, with complete failure around 200 cycles. The thin sponge is delicate and disintegrates under the repeated full chemical reaction of charge and discharge. The lead particles separate from the plates forming micro-short circuits within the battery and resulting in high rates of self-discharge. The net effect of deep cycling is a very rapid decrease in battery efficiency and capacity. Obviously, these batteries are not suited for alternative energy storage or home use.

A recent advance in starting batteries is the sealed or *Maintenance Free* battery. This battery has calcium added to the lead sponge grid to harden it and to reduce water loss due to gassing during charging. The addition of calcium raises the batteries' internal resistance and makes them poor candidates for rapid charging. These types are marginally better for deep cycling than the standard starting battery. Their major advantage is that they require the addition of water much less frequently. The calcium also gives the battery a much lower rate of self-discharge.

Mass production of starting batteries has resulted in very low cost. A 12-volt auto battery of 75 to 100 ampere-hours costs around $70, or about $70 per kilowatt-hour. Even though these batteries are cheaper per kW.-hr., they are more expensive to use in deep cycle applications than the initially more costly true deep cycle types.

In automotive service these batteries last from three to five years. In deep cycle service the starting battery has a life of less than two years—in most cases much less. All estimates of any battery's longevity are dependent on many factors. A good understanding of these factors will help you get the best service from your battery.

LOW ANTIMONY DEEP CYCLE BATTERIES

The demand for a compact, inexpensive deep cycle lead-acid battery has led to the manufacture of a hybrid type. It has some characteristics of both the starting battery and the true deep cycle type. These batteries are labeled "Deep Cycle" but are merely warmed over starting batteries. They contain somewhat thicker plates and usually have some antimony added for additional hardness. This type is easily recognized by its packaging; it uses the smaller automotive case. This low antimony deep cycle battery is once again primarily designed to be compact and inexpensive. It is not suitable for powering large loads over a long period of time. It shares the same deficiencies as the starting battery; it does not respond well to repeated deep cycling. It is designed to meet the requirements of recreational vehicles where space and weight must be held to a minimum.

The cost of an 85 ampere-hour 12-volt low antimony battery is around $85, or about $85 per kilowatt-hour; slightly more expensive than its automotive relative.

In the type of service encountered in recreational vehicles (less than 20 percent discharge before being refilled) these batteries will usually last for between 200 to 400 cycles. If they are cycled to 80 percent or more they will last less than 200 cycles; about the same as a starting battery. The thicker plates with some antimony do indeed add a degree of mechanical strength to the battery. In float (less than ten percent discharge per cycle) service they commonly last between five and ten years; much greater longevity than starting batteries in float service.

TRUE DEEP CYCLE HIGH ANTIMONY BATTERIES

This type of lead-acid battery is designed to have over 80 percent of its energy removed and replaced repeatedly over a period of 5 to 15 years. Although it uses the same chemical reaction as the automotive starting battery, it has little physical resemblance to it. The true deep cycle battery is a highly evolved machine with longevity as the primary function. These batteries are easily recognized by their massive size and weight. Cells of the true deep cycle type are rarely assembled into batteries of over 6 volts because they would be too heavy to move by hand.

The grids in a deep cycle battery are over four times thicker than those in starting batteries, and these grids contain many times the amount of antimony. In all lead-acid batteries the ampere-hour rating, or energy density, is a function of the surface area of the plates exposed for chemical reaction—the more plate area the more powerful the battery. The additional thickness of the plates adds little to the energy density of the deep cycle battery but does add years to their life. These thicker plates are not constructed of sponge lead but of scored sheet lead alloyed with up to 16 percent antimony. The antimony does not enter into the batteries' chemical reaction and is merely along for the ride to provide more strength and longer life to the plates. The combination of thicker gridded plates with more antimony lowers the energy density of the battery, making it larger, heavier and much more costly per kilowatt-hour.

In addition to the plate differences, the deep cycle type comes with a much larger and more rugged battery case. The plates of the cells commonly have 1 to 3 inches of space under them so that particles which have separated from the plates will not cause micro-short circuits on the bottoms of the cells. The tops of the plates are also separated from the top of the battery's case by more distance. This greater "head and foot" room in the battery case gives space for more electrolyte expansion. The increased electrolyte volume within the case provides a reservoir against the water loss common to the charging of deep cycle batteries. The case of a deep cycle battery is not only larger than a starting battery, but it is also designed to be opened and the plates removed and rebuilt. This type of case has an added advantage for the user. The interconnect straps which connect the cells in series within the battery are exposed on top of the case. The interconnect on an auto battery is concealed under the top of the case and is not accessible. The exposed interconnect on the deep cycle battery allows measurement of the voltage of each individual cell. This voltage measurement is valuable in determining when it is time for an equalizing charge.

In the deep cycle battery the primary design goal is long life. In addition to the larger more rugged plates and larger cases, some deep cycle batteries are "wrapped" for longevity. This technique involves wrapping the individual plates with an inert perforated plastic mesh which keeps the lead on the plate longer. The lifetimes of wrapped batteries are 25 to 35 percent longer than those without wraps. The wrapping technique also allows the cells to be more densely packed within the case thus lowering the battery's resistance and raising the energy density.

The most common usage of the true deep cycle battery is for powering electric vehicles. Golf carts, forklifts, and mine tractors are some examples. In this type of service this battery is discharged at high rates to 80 percent or more of its capacity. They are then refilled over a period of 10 or more hours to a state of full charge. The stresses of riding in a vehicle and repeated rapid deep cycling are the reasons for the differences in mechanical construction.

The high antimony deep cycle battery is the most cost effective, commercially available battery that can be used for home storage. The price of a 350 ampere-hour, 12-volt battery is about $400, or $95 per kilowatt-hour. Such a battery will weigh over 250 pounds and contain about 4 1/2 gallons of sulphuric acid.

These batteries can be cycled to 80 percent of their capacity between 1000 and 2000 times. Their lifetime is between 5 and 15 years. As with all batteries, their longevity is also a function of how they are used. Proper cycling techniques, environment, and sizing are necessary to ensure maximum life. Without proper energy management, even the finest of batteries can be ruined in less than a year.

GEL CELLS

This type of lead-acid battery is designed for portability. They are usually small and have a jellied electrolyte within a sealed case. The use of a jellied electrolyte allows the battery to work in any position, even upside down. They are used in aircraft and inside pieces of electronic equipment. Recently they have become available for flashlights and other types of portable electric devices. They are designed to be clean and usable in environments where acid vapors and spills are not acceptable.

These cells are designed to be deep cycled over

a long period of time. They are very strict in their charging requirements being sealed cells. They must not be charged or discharged too rapidly. If energy is cycled through the gel cell too rapidly it will gas, just as in any other lead-acid battery. If gassing occurs in large amounts, it will rupture the seal of the case and ruin the battery. Gel cells are prone to sulphation if they remain in a discharged state for long periods of time. They are also subject to the same self-discharging, or local action, as other lead-acid batteries.

At the current time they are also quite expensive. A 7.5 ampere-hour, 12-volt gel cell costs about $50, or $555 per kilowatt-hour—obviously not cost effective for large storage supplies.

The lifetimes of gel cells vary widely. As basically consumer items, they rarely receive proper charging are commonly abused. One example of such abuse is leaving any battery inside a closed automobile on a hot afternoon. The high temperature greatly accelerates self-discharge and shortens the battery's life. With proper care the gel cell will deliver over 1000 cycles and last for five years or more.

LEAD-ACID BATTERY CHARACTERISTICS

Regardless of the type of physical construction, all lead-acid batteries share the same general characteristics. They all use the same lead-sulphuric acid chemical reaction to store energy and are limited by the nature of this reaction. The operating parameters for all lead-acid batteries are therefore similar.

Voltage

The voltage at the battery terminals is a function of several factors: rate of energy transfer (either charge or discharge), state of charge, temperature, and to some extent, type of cell construction and condition. To further complicate matters, all these factors are interrelated. In order to understand the relationships between them we need to study each separately, but we must always keep

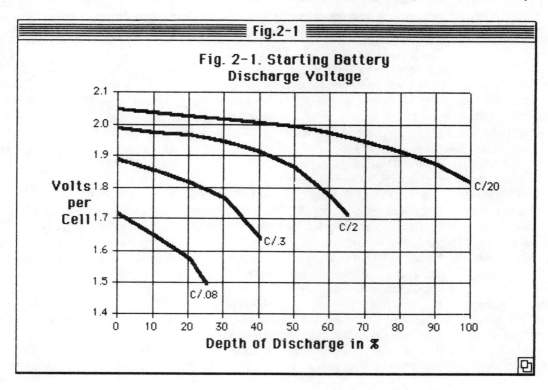

Fig. 2-1. Starting Battery Discharge Voltage

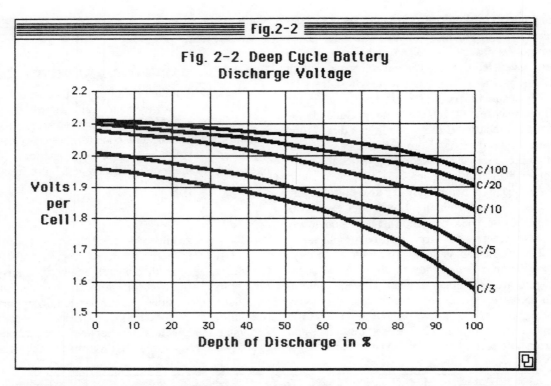

Fig. 2-2. Deep Cycle Battery Discharge Voltage

in mind that they are indeed interrelated and no one factor changes alone. The knowledge of the limits of these factors makes it possible for us to determine when the battery is effectively empty and ready to be recharged and when to stop charging because the battery is full.

As a lead-acid battery is discharged, the voltage of the battery decreases in direct proportion to the rate of discharge. The more load that is applied to the battery, the lower its voltage becomes. In this case the different types of lead-acid batteries display slightly different characteristics. Figure 2-1 gives the data for starting batteries and Fig. 2-2 for true deep cycle batteries. The curves on these graphs represent the voltage of a lead-acid cell as a function of depth of discharge and discharge rate. As can be seen from these graphs, the faster the energy is removed from the battery the lower its voltage becomes. When the battery approaches 100 percent discharge the voltage decreases rapidly to the point where it is so low as to be unusable. This point is called the "knee" of the voltage curve.

Most battery manufacturers specify the discharge cutoff voltage just past this knee. This discharge cutoff voltage is the point beyond which there is little energy left to be withdrawn from the battery. In fact, discharge beyond this point can permanently damage the battery. Experience has shown that a deep cycle battery is most efficiently run when cycled to about 80 percent of its rated capacity.

Temperature

The voltage of the battery is also dependent on the battery's temperature. Since the energy in the battery is stored in a chemical reaction, temperature is an important factor in all battery operation. The lead-acid reaction proceeds most efficiently at about 78° F. As the battery gets colder, the chemical reaction becomes slower and for any given load the battery's voltage is lower. Anyone who has tried to start a car on a cold morning has experienced this effect. Just as too cold is hard on the battery so is too hot. Lead-acid batteries do not last long in continual temperatures over 95° F. These high temperatures, while they don't reduce the voltage very much, do shorten the lifetime of the battery and greatly reduce its efficiency due to self-discharging. Figure 2-3 gives the battery's discharge voltage as a function of temperature and state of charge. The lesson to be learned from this graph is to keep the battery's temperature as close to 78° F. as possible. Batteries kept outside in the winter will not deliver their energy readily, nor efficiently.

Voltage as a Function of State of Charge

One of the most important factors to anyone who is actually using batteries for energy storage is the state of charge of the battery. The most convenient method of determining how much energy is in the battery is by measuring its voltage. Unfortunately, many factors other than state of charge can also affect the battery's voltage. Figure 2-4 gives the rest voltage of a single cell as a function of state of charge. Figure 2-5 gives the same information but for convenience, the voltage is that of a 12-volt battery (i.e., six lead-acid cells in series). These graphs assume the battery is at or near 78° F.; batteries at lower temperatures will show lower voltages. They also assume that the battery is at rest—it is being neither charged nor discharged. If a battery has been recently charged, its voltage will appear higher and it must be allowed to rest for at least 24 hours before the voltage measurement gives an accurate reference for determining state of charge. If a battery is being discharged, it should be allowed to rest at least two hours before making the voltage measurement.

Using voltage measurement over specific gravity for determining state of charge has the large advantage of not having to open the cell to make the measurement. In measuring the specific gravity of the battery, we must open the cells and use a hydrometer to gauge the density of the electrolyte. This procedure can and often does introduce contaminants into the cells. Batteries are chemical

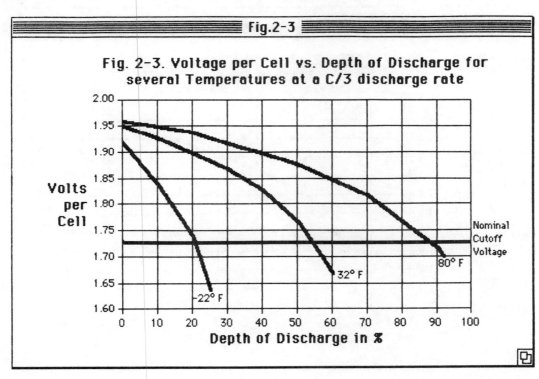

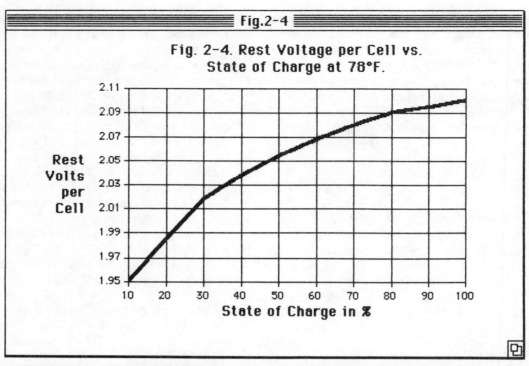

Fig. 2-4. Rest Voltage per Cell vs. State of Charge at 78°F.

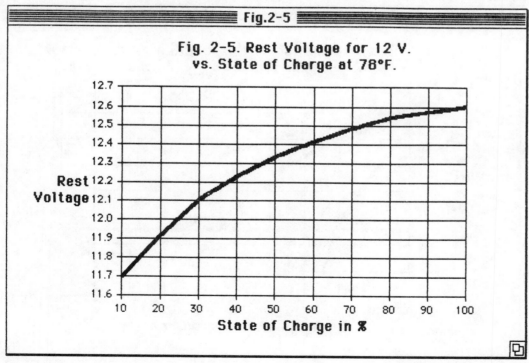

Fig. 2-5. Rest Voltage for 12 V. vs. State of Charge at 78°F.

machines; purity of the reactants is essential. Voltage measurement is easy to perform and does not require that the cells be opened, lessening the chances of contamination. A 12-volt battery has a general voltage operating range between 11.7 and 12.6 volts. Obviously, the voltmeter must be accurate and have the resolution to measure voltage accurately to at least a tenth of a volt.

Specific Gravity as a Function of State of Charge

The state of charge of a lead-acid battery can be determined by specific gravity measurement. When we measure specific gravity with a hydrometer, we are comparing the density of the battery's electrolyte to that of water. In discharge, the sulphuric acid is gradually consumed from the electrolyte and its density goes down. Such a measurement is based on volume; hence it is very temperature dependent. For practical reasons we can only sample the electrolyte directly above the battery's plates. This sample may vary widely from the average of the entire electrolyte due to electrolyte stratification in the battery. Figure 2-6 gives the state of charge as a function of specific gravity for 32°, 55°, and 78° F. If the battery is colder, the specific gravity will measure higher. If the battery is warmer, it will measure lower. Most types of hydrometers are available in temperature compensated models and are certainly worth the extra cost.

Capacity as a Function of Temperature

Temperature plays a key role in determining how the lead-acid battery will perform. Temperature affects the actual capacity of the battery. When a manufacturer rates his batteries in ampere-hours he does so at 78° F. Batteries which are colder have, in fact, less energy to deliver and have less than their rated capacity. Figure 2-7 gives the capacity of the battery as a function of temperature. The main conclusion to be drawn from this information is that if you are going to keep your batteries outside in a cold climate you must

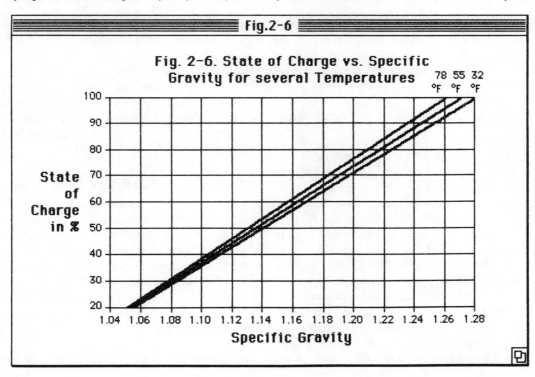

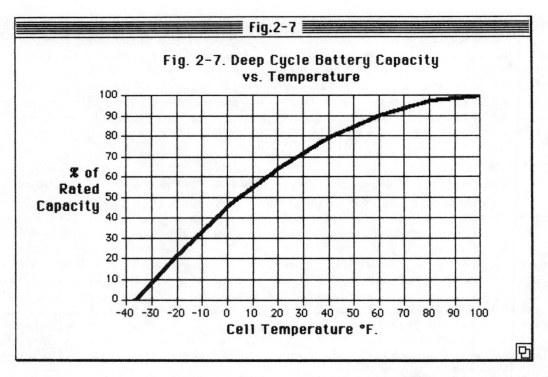

Fig. 2-7. Deep Cycle Battery Capacity vs. Temperature

oversize them in terms of capacity. Batteries that are constantly in the cold for months at a time are prime candidates for sulphation and early replacement. They are also less efficient in their transfer of energy.

Longevity as a Function of Depth of Cycle

One of the main factors affecting the length of a battery's life is how far it is discharged before being refilled. The deeper the cycle the shorter the battery's life. This is especially true for non-deep cycle types of lead-acid cells. It must also be remembered that lead-acid batteries have a limit to their life, no matter the depth of cycle or discharge/charge rate. So the decision is one of mini-max—how to optimize the depth of cycle to give the most efficient power possible. Figure 2-8 gives the battery's life in number of cycles as a function of depth of discharge in percent. This information seems to indicate that if the battery is not discharged deeply it will last very much longer. However, these batteries have a finite lifetime and in practical service even the finest expire of old age before 2,000 cycles. The optimum cycle depth for the true deep cycle batteries is about 80 percent. This figure is the result of a trade-off between battery cycle life vs. depth of cycle and absolute calendar life of the battery. This information combined with an accurate estimation of energy consumption is necessary for proper sizing of battery packs to particular jobs.

Efficiency as a Function of Discharge Rate

Batteries are machines which store electrical energy in chemical reactions. No machines are 100 percent efficient—batteries are no exception. We can get no more energy out of a battery than we have put in it—there is no free lunch here. The lead-acid cells average an efficiency of about 70 percent. The factors and reasons for the energy lost in storing and retrieving are built into the chemical reactions and are covered in the technical section at the end of this chapter. Figure 2-9 shows the relationship between the energy efficiency of the battery

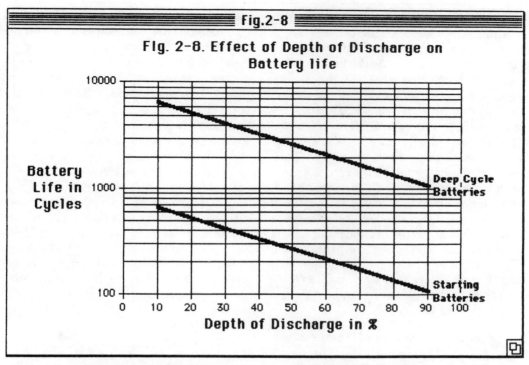

Fig. 2-8. Effect of Depth of Discharge on Battery life

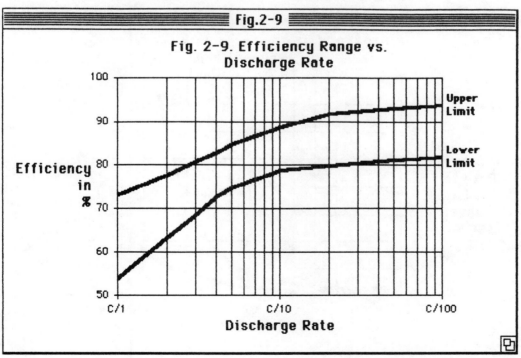

Fig. 2-9. Efficiency Range vs. Discharge Rate

and the rate at which it is discharged. The discharge rate is expressed in terms of the battery's capacity in ampere-hours divided by the number of hours it takes to fully discharge the battery. This is the equation's form—Rate = C/T. C is the capacity expressed in ampere-hours and T is the time it takes to fully discharge the battery in hours. This information indicates that the battery is more efficient if discharged slowly. If we know how much power we need, we can convert this power figure to an average discharge rate. This rate is a fixed and given factor—it's how much energy we need. The only choice we have to make is how much capacity we need from the battery pack. The smaller the pack in ampere-hours, the faster we will discharge it and the less efficient the batteries will be. The relationship between efficiency and discharge rate indicates that we should have as large a pack as possible. However, there are other factors determining capacity.

Local Action as a Function of Temperature

The rate of discharge is not the only factor to affect battery and overall system efficiency. These large batteries are costly. It is a waste of money to have many times the capacity that is needed. These batteries have a distinct calendar lifetime and capacity that is not used is lost to old age anyway. To compound the problem, a lead-acid battery will slowly discharge itself even if no load whatsoever is present.

This process of self-discharging is know as *local action*. Local action is primarily a function of temperature and battery age. Figure 2-10 represents the self-discharge rate in percent of capacity per week as a function of temperature. This graph is based on the performance of deep cycle lead-acid batteries. This information shows that deep cycle batteries kept at 78° F. will lose about 5 percent of their capacity per week. In other words, in a ten-week period the pack will lose about 50 percent of its capacity.

The graph seems to indicate that if we keep our batteries cold, they will discharge themselves at all. This fact is true: however, we will also be unable to retrieve the energy from the battery for our use.

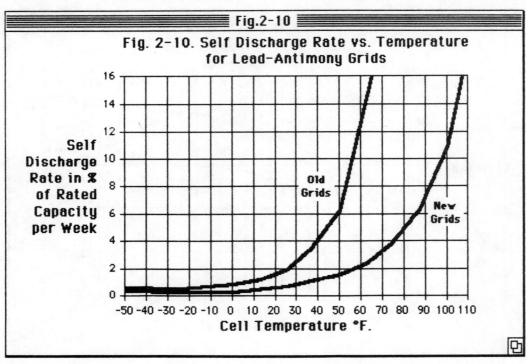

Fig. 2-10. Self Discharge Rate vs. Temperature for Lead-Antimony Grids

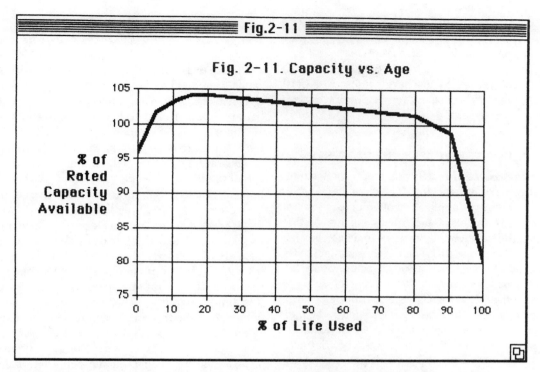

Fig. 2-11. Capacity vs. Age

The same relationship between temperature and the lead-acid chemical reaction that slows down local action also reduces the battery's ability to deliver power to a load.

When all effects are considered, the best temperature for running a lead-acid battery is still around 78° F. However, if a lead-acid battery is to be stored and not cycled for a period of several months or longer, a cold place should be selected. Around 40° F. is best for storage. Make sure the battery is fully charged before storing and that it is warmed up before applying loads to it.

Capacity as a Function of Age

The older a lead-acid battery is the more prone to local action, or self-discharge, it gets. Figure 2-11 shows the relationship between a battery's age and its capacity. In effect, the older a battery gets the less capacity it has. More of the charging energy is lost within the cell and is not available for use. A battery which is eight- to ten-years old may discharge itself over four times faster.

It is not cost effective to maintain a battery pack that is many times larger than what we need. The effects of local action would become significant if the battery has more capacity than we can use in two or three weeks. If the same batteries are over ten-years old, then we have to cycle them weekly just to overcome the effects of self-discharge. The advantage of oversizing the capacity of the batteries so that the rate of discharge is lower is overcome by local action if 80 percent of the capacity is not used in a one- to three-week period.

Proper sizing of the batteries is essential if they are to perform satisfactorily. Too big is as inefficient as too small. The sizing of battery packs to meet specific needs is discussed fully in Chapter 9.

All this data on lead-acid batteries plus practical experience makes possible the formulation of a number of basic rules for efficient battery usage. These rules apply specifically to deep cycle batteries in alternative energy storage service.

1. Keep the temperature of the batteries as close to 78° F. as possible.

2. Keep the batteries clean. If you use a hydrometer, be sure it is totally clean.

3. Use only distilled water to replace the lost electrolyte.

4. Avoid prolonged discharge rates over C/10. Size the battery's capacity for a complete cycle in one to three weeks.

CHARGING THE LEAD-ACID BATTERY

The process of recharging a battery is simply the reverse of discharging. The chemical reaction is reversed by the application of voltage across the battery's terminals. This causes current to flow backwards through the battery—ionizing the electrolyte. The process of recharging is subject to the same factors and limits as discharging. Some factors of a battery's service we can control, some we cannot. Of all the controllable factors during charge, the rate at which the battery is charged is the most important. Proper charging is essential if the batteries are to be efficient and long-lived. Batteries can be rapidly destroyed by continual overcharging or too rapid charging. Batteries that are constantly undercharged will become sulphated and inefficient.

The cardinal rules for charging a lead-acid battery are: charge it as soon as it is empty, and fill it all the way up. Lead-acid batteries which languish about discharged are going to sulphate rapidly and shortly be of little use.

Rate of Charge

How fast the batteries are filled(rate of charge) and at what point they are considered to be "full" are critical factors affecting battery life and efficiency. The two physical measurements necessary for determining rate of charge and state of charge are amperage and voltage.

The rate at which a battery is charged is expressed as the battery's capacity in ampere-hours divided by the time it takes to charge the battery. In equation form this looks like:

Rate of Chart = Capacity/Charge time

$$I = \frac{C}{T}$$

Here is an example for a battery with the capacity of 400 ampere-hours. If this battery was charged at 40 amperese, then the rate is C/10; the battery will be full in 10 hours. 40 amperes times 10 hours equals 400 ampere-hours. A C/20 rate for the 400 ampere-hours battery would be 20 amperes. At this rate the battery will be full in 20 hours. In actual practice, since the battery is less than 100 percent efficient, the duration of the charge must be increased some 20 percent.

Here is how to figure the correct charge rates for any particular lead-acid battery. Take the capacity of the battery in ampere-hours and divide it by 20. This rate in amperes (C/20) is generally the most efficient rate to charge the battery when it is almost empty or almost full, i.e. less than 20 percent or greater than 90 percent state of charge. When a lead-acid battery is either almost empty or almost full its ability to store energy is reduced. This is due to changes in the cell's internal resistance.

Attempting to rapidly charge a battery that is either empty or full causes gassing and increased heating within the battery, greatly shortening the battery's life. Between 20 percent and 90 percent state of charge, the fastest rate of charge is C/10, or the capacity of the battery divided by ten. This is the fastest rate that is efficient to charge a lead-acid battery. The C/10 rate is not as efficient as slower rates; more energy is wasted in heat. At the C/10 rate the battery would be full in 10 hours if it were 100 percent efficient, which it is not. In actual practice the C/10 rate takes about 12 hours to fill the battery.

Battery Voltage while Recharging

As the battery is being charged its voltage begins to gradually rise. By measuring this voltage and by knowing the rate of charge, a rough approximation of the battery's state of charge can be determined. Figure 2-12 gives the charging voltage per cell as a function of state of charge for several

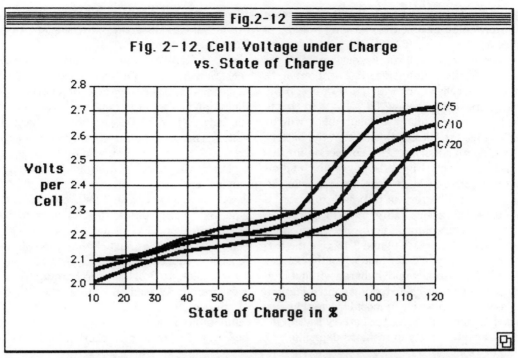

Fig. 2-12. Cell Voltage under Charge vs. State of Charge

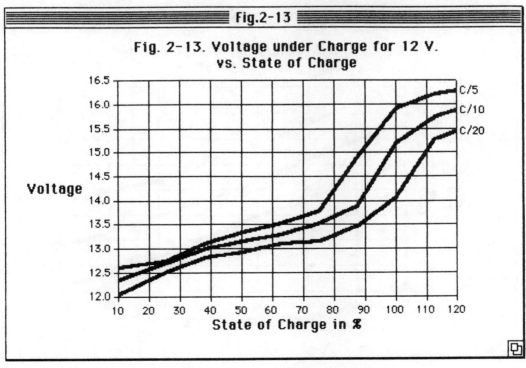

Fig. 2-13. Voltage under Charge for 12 V. vs. State of Charge

charges rates. For convenience, Fig. 2-13 gives the same information except the charging voltage is based on a 12-volt pack (6 lead-acid cells in series).

The problem with all voltage measurements of lead-acid batteries is temperature. The internal resistance of the cells varies with temperature. If the batteries are cold (below 50° F.) and being charged, then the voltage measurement will be higher and lead to the conclusion that the battery is fuller than it actually is. As a correction for this temperature effect, Fig. 2-14 gives the end-of-charge cell voltage (the voltage measurement indicating that the battery is full) as a function of temperature for various charge rates. Figure 2-15 provides the same information, only based on a 12-volt system. Figures 2-14 and 2-15 give the battery user the ability to determine when the battery is full while it is actually under charge without having to open any of the cells, risking contamination.

It is common practice in most charge schemes to limit the upper voltage of the battery pack. This is done in a wide variety of fashions depending on the method of charging. This voltage limit is established so that the batteries are not overcharged and damaged. As the battery fills, the voltage rises. The limitation of voltage causes the current flowing through the battery to gradually decrease, tapering the rate of charge. Voltage limitation is also important if loads are present on the batteries while they are being charged. The voltages across a battery pack under charge may be too high for electronic and other dc appliances on line at the time, resulting in destruction of the appliance.

A general voltage limit for a lead-acid cell under a periodic charge of less than 24 hours in duration is 2.5 volts, or 15 volts for a 12-volt pack. If the cells are under continual charge, the voltage limit must be reduced to 2.2 volts, or 13.2 volts for a 12-volt pack, to prevent overcharging. As with all measurements based on voltage, the temperature of the batteries must be considered. If the temperature is above 90° F., the voltage limit must be lowered. If the temperature is less than 50° F.,

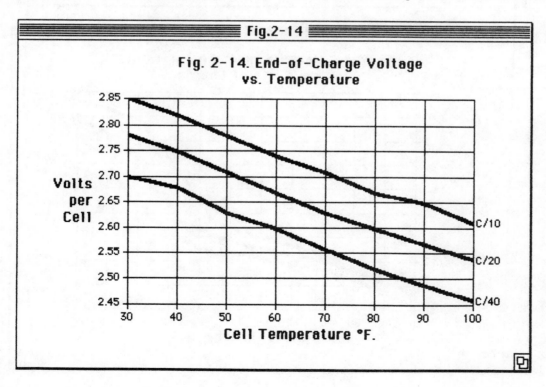

Fig. 2-14. End-of-Charge Voltage vs. Temperature

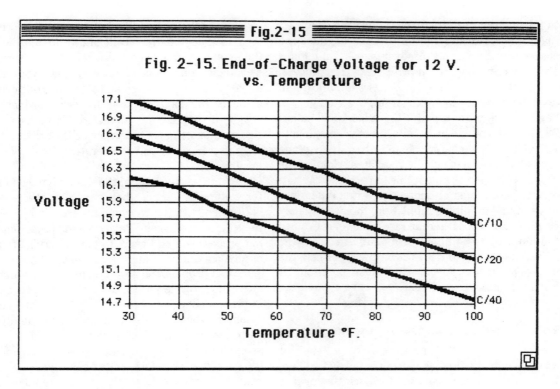

Fig. 2-15. End-of-Charge Voltage for 12 V. vs. Temperature

then the limit must be raised.

Using voltage limitation as the primary control for charging lead-acid batteries is less reliable than using current (amperage) limitation. If voltage is the only limit on the charging process, the battery may be charged far too rapidly when it is empty. The ideal charging process involves the use of current limitation until the battery is almost full and then using voltage limitation to finish the process. Methods and machines for accomplishing this are detailed in Chapter 6.

The foregoing discussion assumes the ability to charge the batteries at will, using constant rates whenever they are empty. Most users of alternative energy systems do not have this ability. They must take power from their systems when it is available and in the quantities that it is available. The "now you see it, now you don't" nature of devices like photovoltaic solar cells and wind machines makes filling the batteries an uncertain process. Proper sizing of the charge source to the battery's capacity is absolutely essential if the system is to work and survive.

A charge source should be capable of delivering at least a C/20 rate to the batteries. It should have some provision for voltage limitation to guard against overcharging and appliance destruction. From the batteries' point of view, every solar and wind system should have a second power source capable of proper battery charging. Such a backup source is essential for equalizing charges and to prevent the batteries from being constantly discharged when the wind is inactive or the weather cloudy for long periods of time.

Let's examine a typical charge on a deep cycle battery. This example gives the optimum method of recharging deep cycle lead-acid batteries. For example, let us consider an empty 300 ampere-hour battery. The charge should begin at a rate of C/20, or 15 amperes for a 300 ampere-hour battery. This charge rate may be increased to C/10, or 30 amperes, after the battery reaches a state of charge of 20 percent or greater. After a period of about 8 hours at the C/10 rate, the battery will be close to full and the rate should be reduced to C/20. Actually the C/20 rate can be used throughout the charge process; its only disadvantage is that it takes about twice as long to complete the charge process.

At the point when the battery is almost full (over 90 percent state of charge), the voltage will begin to rise more rapidly. For a lead-acid cell at 78° F., this voltage point will be between 2.35 and 2.5 volts. The charge should then be voltage regulated instead of current regulated. Holding the charge voltage constant at this time causes the battery to be charged at decreasing rate. When it is full and under voltage regulation, the battery may continue to be charged at rates between C/30 to C/100 for up to 24 hours with no ill effects. If the voltage is limited to 2.2 volts or lower per cell, the charging process can be carried on indefinitely. Such a continuation of charging, even though the battery is full, is called *float charging*.

Battery Maintenance During Recharging

The best time to perform several types of maintenance is when the battery is full and still being charged. This is the time to check the level of the electrolyte in each and every cell. Since the battery has been under charge for hours, the electrolyte is as warm as it is going to get. Filling the battery when it is cold and then charging it will cause electrolyte to overflow as it heats up and expands during the charging process. When the battery is full and being charged it gases, and these bubbles of gas help to evenly distribute the added water throughout the electrolyte. Remember to use only distilled water to replace lost electrolyte.

The second type of maintenance is checking the voltage differences between the individual 2.2-volt lead-acid cells that make up the battery. If the cell voltages differ by more than 0.05 volts, the battery must be given an *equalizing charge*. The equalizing charge is a highly controlled overcharge of the battery. Its function is to bring all the cells in the battery to the same state of complete charge.

To determine if equalization is necessary, measure the voltage across each and every cell in the battery. The absolute value of the voltage measurement is not important; what we are looking for are the differences between cells. Normally a lead-acid battery will show no more than 0.05-volts difference between cells. If one or more cells show a difference of more than 0.05 volts, then equalization is necessary and should be performed at once. Imbalances in the battery will increase rapidly if they are not corrected. The reasons for this imbalance in the different cells of the battery are differences in cell temperatures and rates of local action.

After the checking of the electrolyte level and for voltage differences between cells, the battery is finished with the charging process and is ready to be discharged again. Following the proper charging process will add years to the battery's life and assure maximum system efficiency.

Most problems in batteries used in alternative energy deep cycle service can be traced to repeated improper charging techniques. One of the most common problems is being unable to totally fill the batteries. Batteries charged by wind and solar spend most of their time only partially charged and they hardly ever reach a state of charge of 100 percent. This leads to sulphation and the rapid demise of the batteries. Batteries in these types of services should be equalized monthly.

The Equalizing Charge

When the individual cells in a lead-acid battery begin to show differences in their states of charge, it is time to run an equalizing charge. These internal differences result from differences in temperature and rates of local action between the cells. These differences can be measured in either voltage or specific gravity. The only cure for these inequalities is a controlled overcharge of all cells—the equalizing charge.

In order to equalize a battery it must first be fully charged. The state of charge of the battery may be determined from Fig. 2-12. The voltage limit is then raised to 2.75 volts for a single cell, corresponding to 16.5 volts for a 12-volt system. This voltage level is dependent on temperature and will have to be raised for cold batteries. The battery charge is continued at the C/20 rate for 6 to 10 hours, or until all inequalities in the cells are eliminated. The equalizing charge is actually a constant current overcharge. Equalization should not be carried out at rates greater than C/20.

Caution should be observed as the system voltages, while equalizing, are quite high. A 12-volt system may reach over 16.5 volts during equalization, much too high for many 12-volt appliances. Be sure to turn off all voltage sensitive devices during the equalization charge.

During the equalization process, the battery will gas profusely. Care should be taken to provide adequate ventilation during all charging processes, especially during equalization. The gases (hydrogen and oxygen) emitted from a lead-acid battery during charging are explosive. Keep flames and sparks well away from the battery during all charging. The more rapid gassing during equalization results in increased water consumption of the cells—keep your eye on the electrolyte levels.

If equalization is not periodically carried out, the weaker cells within the battery pack will become overdischarged and wear out quickly. Lead-acid deep cycle batteries should be equalized every five to ten cycles or monthly, whichever comes first. The same equalization is necessary for lead-acid batteries in float or shallow cycle service.

Equalization can be used to restore some of the capacity of aging batteries. It is also effective in dealing with the sulphation which results from sitting discharged for extended periods of time. If the batteries are in very poor condition, they may take up to five equalizing charges to restore them to service. The equalizing charge will not restore totally worn out batteries, but it will help us keep our batteries working efficiently for as long as possible.

Electrolyte Stratification

Another problem solved during equalization is the stratification of the electrolyte. Under charge and discharge, the amount of sulphuric acid present in the electrolyte changes. Since sulphuric acid is more dense than water, the more concentrated and highly charged portion of the electrolyte collects in the lower parts of the cells. This stratification of the electrolyte leads to premature wear within the cell. Some types of large lead-acid stationary storage cells have pumps just to circulate the electrolyte and prevent stratification. The rapid evolution of gases during the equalizing charge causes the electrolyte in the cells to bubble and be mixed, also preventing stratification.

The advantages of equalization are so great that users of solar cells and wind powered generators should consider the construction of a charging machine just for equalizing charges. Chapter 6 contains a discussion of such a machine that is cheap and easy to build.

DISCHARGING THE LEAD-ACID BATTERY

As a lead-acid battery is discharged the sulphuric acid is consumed from the electrolyte. The voltage of the battery drops. The faster the battery is discharged (higher rate of discharge), the faster the sulphuric acid is consumed and the faster the voltage drops. The voltage of the battery pack is critically important to users of battery-stored energy. All loads and appliances are designed to operate over a voltage range. If the battery voltage falls below this range, the device will not operate properly if at all.

Discharge Rate

If a battery is fully charged and at room temperature, the single most important factor determining battery voltage is the rate of discharge. The rate of discharge is proportional to the battery's capacity. A 10-ampere load is a C/10 discharge rate for a 100 ampere-hour battery, while the same 10-ampere load is a C/40 rate for a 400 ampere-hour battery. For the same load, larger batteries have slower discharge rates in relation to their capacity.

When considering the prolonged discharge of lead-acid deep cycle batteries, the discharge rate should be less than C/20. Another way of saying this is that the battery would be empty after no less than 20 hours of continuous use. In actual practice, the loads are intermittent. The battery has its stored energy discharged in periods of use and periods of rest. If the voltage is to remain acceptable to all loads, a load of C/20 should be considered a maximum for continuous usage. Loads much greater than C/20 can be handled by the batteries as long as they are not long in duration. Sizing the battery pack to handle at least a C/20 load will assure long life and adequate performance. See Figs. 2-1, 2-2, and 2-9 for this information in graphical form.

What this information translates to in actual practice is this—don't ask for energy faster than the batteries can efficiently deliver it. If the maximum continuous load expected is 50 amperes, then an appropriately sized battery pack would be 1000 ampere-hours. Such a pack could power the 50-ampere load for 20 hours. Since the 50-ampere load will probably only be used intermittently, the batteries will actually last much longer than 20 hours. Even considering many other small (under 10-ampere) loads, a 1000 ampere-hour battery could last an energy conscious home more than a week. Chapter 9 discusses the techniques of load estimation and battery sizing at length.

Self-Discharge or Local Action

Another factor to be considered in relation to discharge is local action. The tendency of all lead-acid batteries to discharge themselves while just sitting there can have a definite effect on battery performance in the discharge cycle. Figure 2-10 points out the relationship between local action and temperature. Battery age and temperature are the two most important factors in determining self-discharge rates. In practice, this information tells us that it is inefficient to store energy in lead-acid batteries for extended periods of time. The average deep cycle lead-acid cell loses about 5 percent of its energy over a week's time. This energy is lost from the battery whether we use it or not. If energy is to be stored for longer than three weeks, the local action of the battery becomes a significant portion (15 percent to 18 percent) of the total energy transferred. This problem gets worse as the batteries age. Again, this problem can be solved by proper battery sizing and proper energy management.

It is inefficient to store more energy in a deep cycle lead-acid battery than can be used in a one- to three-week period. Battery packs which have more capacity than is needed in this period will lose appreciable amounts of energy due to local action. The solution is simple—don't buy more battery capacity than is needed.

The efficiency picture for oversized battery packs grows even darker if one also considers the charge cycle. If the batteries are to last, then they must be filled totally and equalized periodically. The amount of energy needed to refill the pack is dependent on the size and state of charge of the pack. The larger the capacity of the battery, the more energy it takes to fill it and to equalize it. So it is a case of having to replace the energy lost from local action even though this energy cannot be actually used.

The age of the batteries also affects the rate of local action. As the battery pack ages, this rate of self-discharge increases and the oversized battery pack becomes even more inefficient. Batteries which are not totally filled age faster than those which are properly charged. A combination of all these factors leads to much more in-efficiency for oversized battery packs. This inefficiency increases rapidly as the batteries age.

The single most important factor, excluding temperature, that affects discharge efficiency is the discharge rate, which is a function of the battery's capacity. Size it big enough to meet the need, but not so big that it becomes inefficient.

Depth of Discharge

The one remaining factor in the discharge cycle over which we have control is the depth of discharge. How deeply a battery is cycled plays a large part in overall system efficiency. Here the picture is complicated with factors that have nothing to do with the actual energy transfer characteristics of the battery itself. The expense and availability of the energy used for charging is such a critical factor. The cost and finite lifetime of the batteries is another.

In general, if we have the ability to fill the batteries at will, then a depth of discharge of 80 percent is the most efficient overall for deep cycle lead-acid batteries. This figure is derived from many interrelated factors and much practical experience.

LEAD-ACID BATTERY TECHNICAL DATA

All current types of batteries are chemical engines. A thorough understanding of the electrochemical reactions taking place within the battery will help us to use batteries more efficiently. Fortunately, the reactions of lead-acid cells are very basic and simple. The processes are understandable to anyone who managed to stay awake during high school chemistry or physics.

Chemical Composition

The positive plates (anodes) within the lead-acid battery are made of lead dioxide (PbO_2). The negative plates (cathodes) are constructed of lead (Pb). The electrolyte is a dilute solution of sulphuric acid (H_2SO_4) and water. In the charged state the electrolyte exists as ions or charged molecules. Both electrodes of the battery are completely immersed in this electrolyte. The reversible chemical reaction between the plates and the electrolyte allows the storage and retrieval of energy from the battery.

The voltage produced across a single lead-acid cell is a function of the electrochemical reaction between the active constituents of the battery. All lead-sulphuric acid reactions proceed at about 2 volts. This is a given factor. If more voltage is needed, then more cells must be added in series. The physical size of the cell is variable and determines the amount of current, at 2 volts, available from the battery. In other words, the more massive the cell the greater its capacity in ampere-hours. No matter how large the single cell is, its voltage will still be around 2 volts.

Discharge Reactions

When a lead-acid battery is being discharged, the active materials of both electrodes are changed into lead sulphate ($PbSO_4$). The sulphuric acid is gradually consumed from the electrolyte. The discharge chemical equations for the anode and cathode follow:

DISCHARGE

Anode: $PbO_2 + 4H^+ + SO_4^= + 2e^- \rightarrow PbSO_4 + 2H_2O$

Cathode: $Pb + SO_4^= - 2e^- \rightarrow PbSO_4$

As the battery is discharged, all the electrodes gradually become plated with lead sulphate ($PbSO_4$) which is an electrical insulator. It will not conduct current. The $SO_4^=$ (sulphate) ions are gradually consumed from the electrolyte and are bonded to the plates to form $PbSO_4$ (lead sulphate). This reaction releases two electrons at the cathode for every $SO_4^=$ radical which is bonded to the plates. This release of free electrons at the cathode from the electrochemical reaction is the source of the battery's power.

During discharge the area of the plates available for reaction decreases as the surface of the plates becomes covered with the insulative lead sulphate. This decrease in the active area results in a sharp rise of the cell's internal resistance and a sharp drop in the voltage of the cell. Eventually the plates have no more area available for chemical reaction as the sulphate ions are consumed from the electrolyte and are chemically combined with the lead on the plates. It is not possible to remove any more energy from the battery. At this point the battery is said to be fully discharged.

Actually the process of discharging is terminated before all of the sulphate ions are consumed from the electrolyte. The ratings of battery manufacturers are based on the actual usable energy, which is much lower than the calculated energy of the battery using the masses of the reactants as a basis. Commercially available batteries are rated between 15 to 40 percent of their theoretical electrochemical capacity.

Charge Reactions

The charging process is the reverse of the

discharging process. During the charging process, a current (flow of electrons) is forced through the battery in the opposite direction by the application of voltage across the battery's terminals. The reversal of the electronic flow within the battery causes the chemical bond between the lead and the sulphate ions to be broken, and the sulphate ion is released into the electrolyte solution. The charge equations for the lead-acid battery are as follows:

CHARGING

Anode: $PbSO_4 + 2H_2O - 2e^- \rightarrow PbO_2 + 4H^+ + SO_4^=$

Cathode: $PbSO_4 + 2e^- \rightarrow Pb + SO_4^=$

When all the sulphate ions have been removed from the plates and are in solution with the electrolyte, the battery is said to be charged. In actual practice all of the ions cannot really be removed from the plates. Some continue to remain bonded to the plates in the form of lead sulphate. The inability of the charging process to remove all the sulphate ions bonded to the plates is one cause of the battery's finite lifetime. In time the plate area available for reaction becomes smaller and smaller as more and more sulphate ions cannot be kicked free of the plates. Such a battery is said to be *sulphated* and suffers from *sulphation*.

Sulphation

Experience has shown that the longer the ions stay bonded to the plates, the more difficult they are to dislodge with the charging process. The equalizing charge ensures that the inevitable process of sulphation is delayed as long as possible.

The nemesis of sulphation is a phenomenon that is not totally understood. Perhaps if it were, we'd have batteries which would last indefinitely. One theory of sulphation is that the nature of the bonding of the sulphate ion to the lead plates changes its character over a period of time. In the normal charge and discharge cycles, the lead and the sulphate enter into ionic bonding, thereby exchanging electrons.

It is also possible for the lead to form *covalent bonds* with the sulphate ion; in such a case the electrons are not exchanged but shared between the lead and sulphate ions. The covalent bonds evolve slowly from the already ionically bonded lead sulphate. The covalent bonds have a much lower energy level than the ionic bonds. Covalent bonds are poor conductors of electricity. The sulphate ions which are covalently bonded require much more energy to free them from the lead on the plates. Such conditions would render the covalently bonded sulphate ions inaccessible under normal charging techniques. The equalizing charge is our only weapon against sulphation. Regular equalizing charges keeps the bonding in the ionic form as long as possible.

Electrolyte

The condition of the battery's electrolyte is of critical importance to the battery's operation. Table 2-1 is a table giving the physical properties of sulphuric acid solutions. Since the measured specific gravity of the sulphuric acid solution is dependent on its temperature, the table contains factors for correction based on temperature. The data presented on the table is calculated for 59° F. (15° C.). The actual specific gravity can be calculated by the equation:

$$SG = SG_{59°F.} + c(59 - t)$$

Where:

SG = specific gravity at temperature t
$SG_{59°F.}$ = specific gravity at 59° F. (available from the table)
c = temperature coefficient per °F.
t = temperature of the solution in °F.

As the battery is discharged the specific gravity of the electrolyte becomes increasingly lower. The freezing temperature of the electrolyte rises at the same time. Freezing of the electrolyte within a lead-acid battery can result in permanent cell damage. The maintenance of a level of specific gravity high enough to prevent freezing is essential.

Table 2-1. Properties of Sulphuric Acid Solutions

Specific Gravity At 59°F.	Temperature Coefficient Per °F.	H$_2$SO$_4$ % Concentration Weight	H$_2$SO$_4$ % Concentration Volume	Freezing Point °F.
1.050	0.00018	7.3	4.2	26
1.060	0.00020	8.7	5.0	
1.070	0.00022	10.1	5.9	
1.080	0.00024	11.5	6.7	
1.090	0.00026	12.9	7.6	
1.100	0.00027	14.3	8.5	18
1.110	0.00028	15.7	9.5	
1.120	0.00029	17.0	10.3	
1.130	0.00031	18.3	11.2	
1.140	0.00032	19.6	12.1	
1.150	0.00033	20.9	13.0	5
1.160	0.00034	22.1	13.9	
1.170	0.00035	23.4	14.9	
1.180	0.00036	24.7	15.8	
1.190	0.00037	25.9	16.7	
1.200	0.00038	27.2	17.7	-17
1.210	0.00038	28.4	18.7	
1.220	0.00039	29.6	19.6	
1.230	0.00039	30.8	20.6	
1.240	0.00040	32.0	21.6	
1.250	0.00040	33.4	22.6	-61
1.260	0.00040	34.4	23.6	
1.270	0.00041	35.6	24.6	
1.280	0.00041	36.8	25.6	
1.290	0.00041	38.0	26.6	
1.300	0.00042	39.1	27.6	-95

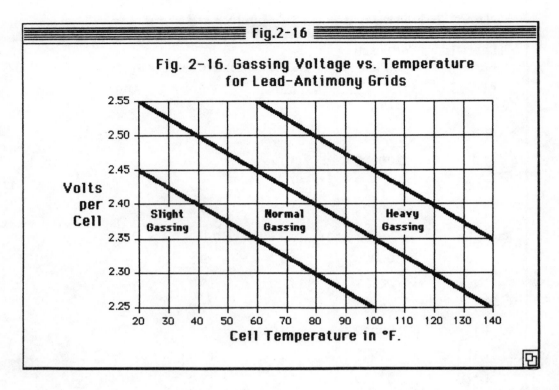

Fig. 2-16. Gassing Voltage vs. Temperature for Lead-Antimony Grids

The chemical composition of the electrolyte is another critical factor affecting battery performance. The reaction within the cells proceeds on a molecular level. If sulphate ions or lead ions become bonded to anything other than each other, they are removed from the energy storage process. Any contaminants introduced into the cell will reduce its capacity. This is especially true of metallic impurities often dissolved in water. Keep the insides of your batteries pure. Use only distilled water to replenish the electrolyte. If a hydrometer is in use, be sure it is totally clean.

Gassing

As the lead-acid cell is charged, gases evolve on the plates. These gases, mostly oxygen and hydrogen, form when the cell voltage exceeds 2.45 volts. Very small amounts of the poisonous gases stibine (SbH_3) and arsine (AsH_3) are produced in the final stages of the charging process. The oxygen and hydrogen gases mostly combine to form water vapor which is emitted from the vents in the cells. This emission of water vapor is the primary cause of water loss in lead-acid batteries. Due to the evolution of these gases during the charge process, the lead-acid cell should only be used where good ventilation is possible. Hydrogen is explosive and dangerous. Unfortunately in order to equalize a lead-acid cell we must raise the voltage over 2.45 volts, so gassing is an unavoidable fact of battery usage. Figure 2-16 gives the gassing voltage as a function of temperature.

Efficiency

The efficiency of a lead-acid battery is determined by the product of the voltage and ampere-hour efficiencies. The ampere-hour efficiency is measured using the current input/output as a basis for calculation. The voltage efficiency is determined using voltage measurements as the standard. The product of these two types of measurements yields a rough figure for the watt-hour efficiency of the battery. The ampere-hour efficiency is primarily a function of the amount of gassing that takes place, as well as the capacity lost due to local action. Voltage efficiency is a function of the charge or discharge rate and the battery's temperature.

Regardless of the approach used to measure the efficiency of the battery, the major area of loss is heat. Energy lost to heat in the cycling of lead-acid batteries comes in two forms. The first is loss due to the inherent internal electrical resistance of the battery. This is called the *I^2R loss*. This factor may be calculated by squaring the current in amperes and multiplying it by the internal resistance of the battery. This factor is exponentially proportional to the amount of energy cycled through the battery per unit time. The greater the charge or discharge rate, the greater the energy that is lost in the irreversible I^2R component. The internal resistance of the cells should be kept as low as possible by keeping the battery at proper operating temperatures.

The second component of heat loss is reversible. It is the heat of reaction between the ions of the battery. When the battery is being charged, the chemical reaction within the battery releases heat; it is said to be *exothermic*. During the discharge cycle the chemical reaction within the battery absorbs heat and is said to be *endothermic*. This emission and absorption of heat is inherent in the chemical reaction between the lead and the sulphuric acid. There is nothing a battery user can do to minimize this factor.

The I^2R component is irreversible, while the heat of reaction component is reversible. During the charge cycle the heat of reaction is exothermic and so is the I^2R component. The battery temperature rises—all this heat is energy lost and is the primary inefficiency in the lead-acid energy transfer process. During the discharge cycle the heat of reaction is endothermic and balances the exothermic I^2R component. What this amounts to is that the bat-

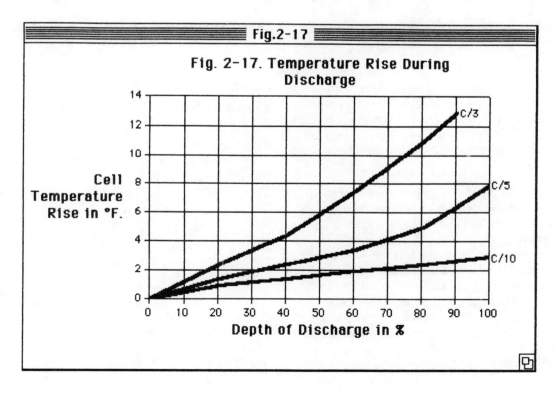

Fig. 2-17. Temperature Rise During Discharge

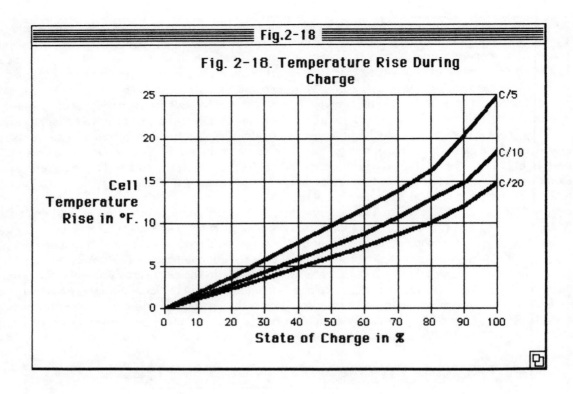

Fig. 2-18. Temperature Rise During Charge

teries seem to get hot when they are charged and not when they are discharged. In either case there is energy lost, whether it is readily apparent or not, to heat in the cycling process. These losses are inherent in the storage of energy in the lead-sulphuric acid reaction.

Figures 2-17 and 2-18 give the relationship between temperature rise and charge/discharge rate. These graphs assume that the batteries are not transferring thermal energy to their surroundings. This assumption that the reaction is *adiabatic* is necessary to minimize the number of factors to be considered. In actual service, the batteries do lose their heat to their surroundings. These graphs are presented as an indication of when and where the primary inefficiencies occur. Figure 2-19 illustrates the electrochemical reactions that take place in the lead-acid cell.

LISTING OF LEAD-ACID DEEP CYCLE BATTERY MANUFACTURERS

Trojan Batteries Inc.
1395 Evans Avenue
San Francisco, CA 94124
415-825-2600

C&D Batteries
3043 Walton Road
Plymouth Meeting, PA 19462
215-828-9000

Gates Energy Products, Inc.
1050 South Broadway
Denver, CO 80217
303-744-4806

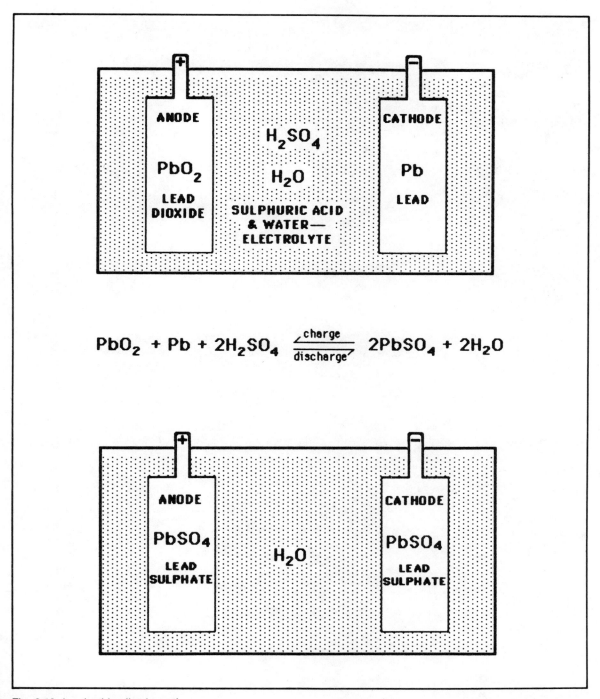

Fig. 2-19. Lead-acid cell schematic.

Exide Power Systems
101 Gibraltar Road
Horsham, PA 19044
215-674-9500

Globe-Union, Inc.
5757 North Green Bay Avenue
Milwaukee, WI 53201
414-228-1200

Gould, Inc.
467 Calhoun Street
Trenton, NJ 08607
609-392-3111

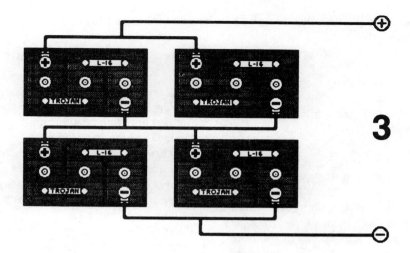

Nickel-Cadmium Batteries

The use of the nickel-cadmium electrochemical couple began around 1900. Nickel-cadmium batteries have been in use in Europe since the 1920s. They became common in the United States in the 1950s.

The nickel-cadmium reaction is very different from the lead-acid reaction. The nickel-cadmium cell is an alkaline battery. Its electrolyte is a base, not an acid. Unlike the lead-acid battery, the electrolyte does not enter into chemical changes with the active materials. The electrolyte (potassium hydroxide in a water solution) acts only as a transfer medium for electrons.

As in the case of the lead-acid battery, the nickel-cadmium battery has evolved into many different forms to meet a variety of applications. Nickel-cadmium batteries are commonly found in small portable appliances. They are less common in large storage supplies because of their expense.

All nickel-cadmium batteries are based on the same chemical reaction. They differ widely, however, in physical construction. These differences in the cell construction result in different operating characteristics for each type of nickel-cadmium cell.

SINTERED PLATE NI-CADS (VENTED AND SEALED)

The sealed sintered plate nickel-cadmium cell is used in many consumer cordless appliances. Almost everyone owns some appliance that is powered by these small rechargeable wonders.

Physical Construction

The sintered plate nickel-cadmium cell is constructed of nickel support plates. The nickel plates are impregnated with the active materials in a powdered form. These plates are then formed and assembled into cells. The sintered plate ni-cad is available in cylindrical, rectangular, and button style cases. These cases are made of nickel plated steel.

The use of powdered active materials allows for very simple manufacturing and low cost. Powders have very large surface areas in relation to their mass. The increased surface areas of the reactants lowers the internal resistance of the cell. The internal resistance of the sintered plate ni-cad is very low in comparison with other types of ni-cads. Even cells of small capacities are capable of high rates of discharge for short periods of time. However, the powdered nature of the reactants makes the cell rather delicate in comparison to the pocket plate ni-cad.

Sintered plate ni-cads are available in two basic types—vented and sealed. The sealed cells are essentially pressurized units. They can withstand over 100 pounds per square inch of internal pressure. Being sealed cells, they will operate in any orientation, even upside down. They do, however, contain a *one-shot* safety vent in case the battery is massively overcharged and gases. The safety vent prevents the cell from exploding or rupturing. If the safety vent is used, the cell's electrolyte quickly dries out and the cell is effectively ruined. The sealed cell also uses a paste electrolyte rather than a liquid one.

The vented sintered plate ni-cad is far less common than the sealed type. In this type of cell the electrolyte is liquid and the cell has a vent in its top. This vent allows the cell to *breathe* at relatively low pressures, about 10 pounds per square inch over atmospheric pressure. It also allows addition of water to the electrolyte should any be lost from gassing. The vent and the liquid electrolyte in this type of ni-cad restrict them to vertical operation only. If a vented cell is turned upside down, the electrolyte will run out and quickly corrode its surroundings. Vented cells with liquid electrolytes are used in large stationary storage systems and not in portable equipment.

The big advantage of the sintered plate nickel-cadmium cell is its power density. A large amount of energy can be packed into a very small cheap portable package. The sintered plate ni-cad is very suitable for powering portable electric appliances. This applies solely to the sealed type, as the vented type is only capable of vertical operation.

The sealed sintered plate ni-cad is similar to the lead-acid auto battery in its design requirements. It is constructed to give as much power as possible in the smallest, lightest, and cheapest package. It is a consumer item and, as such, it is not designed for longevity or ruggedness, but for low cost.

Sintered Plate Ni-Cad Applications

The sealed sintered plate ni-cad is very common in all types of portable rechargeable equipment. Use of this type of battery has made possible many rechargeable portable appliances which did not exist 15 years ago. Radios, televisions, electric razors, drills, flashlights, and cassette tape players are but a few examples. The ability of the sintered plate ni-cad to deliver large amounts of current for short periods of time makes them ideal to power portable electric motors.

The battery packs within these rechargeable appliances usually consist of several sealed sintered plate ni-cads in series. The cells range in size from AAA to D sized packages. The charging unit is usually run on house power (120 Vac). The charging unit may be contained within the device or external to it.

Available Capacities

The sealed sintered plate ni-cad is commercially available in a large number of sizes. Capacities vary from 0.1 ampere-hours to 20 ampere-hours. Most equipment uses the standard size cells. A AA-sized ni-cad has a capacity of about 0.5 ampere-hours. The C-sized ni-cad has a capacity of between 1 and 2.2 ampere-hours. The D-sized cell has a capacity of between 1.2 and 4 ampere-hours. All ni-cad cells regardless of size have an output voltage of around 1.25 volts.

The variations in capacity within each standard size are due to the actual amount of active material used in the construction of the cell. Some manufacturers simply put a C-sized cell into the larger D case and call it a D cell. Consumers should beware and insist on the largest capacity available in any particular standard size cell. A good rough test of the capacity of any ni-cad cell is its weight. The 4 ampere-hour version of the D cell weighs over twice as much as the 1.2 ampere-hour version.

The vented sintered plate battery is very rare as vented cells are mostly used in stationary storage applications. In these types of systems longevity is a critical factor. The sintered plate type is definitely inferior to the pocket plate type in terms of cell longevity. The vented sintered plate ni-cad is not commercially produced in large capacities.

Cost

Costs of sealed sintered plate ni-cads, being consumer items, differ widely at the retail level. Generally, the larger the capacity of the cell the less expensive per kilowatt-hour the cell is. The standard size small cells are priced competitively by most manufacturers. A AAA-sized cell costs about $2.20, or $9,780 per kilowatt-hour. A AA-sized cell costs about $2.25, or $3,600 per kilowatt-hour. A C-sized cell costs about $5.30, or $2,355 per kilowatt-hour. A D-sized cell costs about $9.80, or $1,960 per kilowatt-hour.

At these prices, the batteries are attractive for portable gear but they are definitely not cost effective for large capacity storage. Even the largest capacity sintered plate ni-cads (about 20 ampere-hours) cost in the neighborhood of $900 per kilowatt-hour.

Longevity

The sintered plate nickel-cadmium cell has a lifetime of up to 500 cycles. Its calendar lifetime is about ten years. Calendar lifetime assumes that the battery is in float service and not cycled. In most cases the cycle lifetime is the limiting factor.

Since most batteries are improperly charged, it is common to see them fail after less than 100 cycles. The sections on battery characteristics in this chapter deal with optimizing life through proper cycling. Chapter 6 describes a machine which will charge these small ni-cads properly.

POCKET PLATE NI-CADS (VENTED AND SEALED)

The pocket plate nickel-cadmium cell is commonly used in both its vented and sealed types. The vented form is usually encountered in large storage batteries. The sealed form is available in standard dry cell packages and is used in the better grade of rechargeable portable equipment.

Physical Construction

The nickel plated steel support plates of the pocket plate are covered with perforated cavities. The small web-like "pockets" contain the cell's active materials. The active materials in the pocket plate are highly constrained by the pockets from thermal expansion and contraction. This type of construction results in greatly increased life by reducing the loss of the active materials from the cell's grid.

These spheres of active material are hundreds of times larger in size than the powders found in the sintered plate type. The larger spherical form has less surface area per unit weight than the powders. The decrease in the ratio of surface area to the weight of the reactants changes the operating characteristics of the cell. The pocket plate type has more internal resistance and less energy density than the sintered plate type. It is not capable of extremely high discharge rates in relation to its capacity. While the pocket plate cannot transfer energy as rapidly as the sintered plate, both types of ni-cads are superior to the lead-acid reaction in rapid energy transfer.

There are certain advantages to the pocket plate. The structural integrity of the pocket style gives the cell greatly increased longevity. The ruggedness of the mechanical construction makes the cell able to withstand much higher thermal stress. The pocket plate ni-cad, either sealed or vented, is remarkably resistant to damage from the heat of overcharging. The pocket plate ni-cad is also less prone to self-discharge at high temperatures (over 110° F.) than is the sintered plate cell.

The sealed pocket plate ni-cad is usually manufactured in small standard sized cells. It is designed to be portable and has a paste electrolyte. It contains a safety vent to relieve internal cell pressure in the event of gross overcharging. If this safety vent is used, the cell will rapidly dry out and become useless.

The vented type of pocket plate ni-cad cell is used in large energy storage systems. Its physical appearance is very similar to the lead-acid deep cycle battery. The electrolyte exists in a liquid state. The nickel grids are housed in large rectangular cases with the output terminals and a removable vent cover on top of the cell. The case may be plastic or nickel plated steel.

Most vented pocket plate ni-cads are cased as single cells. They are quite massive and can weigh several hundred pounds. These cells are not designed to be portable, but to provide large capacity energy storage in stationary applications.

Pocket Plate Ni-Cad Applications

The sealed pocket plates are used in the same applications as their sintered plate brothers—small rechargeable portable appliances. Manufacturers who are attempting to market high quality products use the sealed pocket plate ni-cad in place of the sintered plate type. Professional video cameras, tape recorders, and computer memory backup batteries are examples of pocket plate usage. In general, the pocket plate cell can be used to replace the sintered plate type in applications demanding increased reliability.

For the consumer, it is difficult to recognize what type of ni-cad he is buying. Most rechargeable equipment using ni-cads does not specify the type of cell being used. The pocket plate cell's ability to be repeatedly and rapidly charged is one clue to the the type of the cells in an appliance.

If a device can be recharged in three hours or less, it is possible that the ni-cads contained within it are of the pocket plate type. If the device can be recharged within 1 hour, the battery is definitely composed of pocket plate cells. Pocket plate cells weigh more than the sintered plate type for the same cell size. This increased density of the cells is also a clue to the type of cell being used. The extra initial cost incurred is more than balanced by the increased longevity of the pocket plate type of ni-cad.

The vented style of pocket plate ni-cad is used almost exclusively in large stationary storage systems. These types are manufactured to meet the needs of home energy use, emergency power, and load leveling by electrical utility companies. The vented pocket plate ni-cad is highly optimized to perform efficiently and for long periods of time. In constantly cold climates, they offer greater cost effectiveness than the lead-acid deep cycle types due to their wider range of operating temperatures. The relatively constant output voltage of the vented pocket plate ni-cad is highly desirable in alternative energy service. This type of storage cell offers the lowest self-discharge rate of any affordable battery.

Available Capacities

Sealed pocket plate ni-cads are commercially available in capacities ranging from 0.8 to 30 ampere-hours. They are manufactured in the larger dry cell packages. Use of the C- and D-size cells is common in portable professional video equipment. The smaller dry cell packages are not available in the pocket plate design.

Vented pocket plate cells are available in capacities between 5 and 1,400 ampere-hours. This type of ni-cad cell is the only kind manufactured with sufficient capacities for alternative energy applications. The available capacities of these large storage cells are dependent on the discharge rate and operating temperature.

Cost

The sealed pocket plate is more expensive per kilowatt-hour than the sealed sintered plate. The pocket plate is more complex to manufacture. A C-sized package of 2.2 ampere-hours costs about $9.90 or $3,600 per kilowatt-hour. The D-sized cells of 4 ampere-hours costs about $13.80 or $2,760 per kilowatt-hour. The sealed pocket plate, while initially more expensive, is more cost effective if used for several years.

The vented pocket plate ni-cads cost from $400 to $1,200 per kilowatt-hour. An alternative energy battery of 12 volts at 700 ampere-hours costs in the neighborhood of $4,500 to $6,000. These ni-cads are about four to ten times more expensive per kilowatt-hour than a comparable lead-acid battery.

Longevity

The pocket plate design has the greatest longevity of any type of nickel-cadmium cell. Sealed versions have a cycle life between 500 and 1,000 cycles. Their calendar lifetime is around 15 years. Calendar lifetime assumes the cells are in float service and not cycled.

The vented pocket plate has a cycle lifetime of 1,500 to 2,000 cycles. Its calendar lifetime is about 20 years. These figures should be considered as maximums which can only be reached through careful use. Improper cycling or maintenance can greatly reduce the lifetime of any type of battery.

NI-CAD BATTERY CHARACTERISTICS

All nickel-cadmium batteries use the same chemical reactions to store energy. The limitations of these chemical reactions determine the operating parameters of the battery. As such, all types of ni-cads share the same general operating characteristics. The minor differences that do exist between ni-cads are due to differences in mechanical construction.

Voltage

At moderate discharge rates (C/5 to C/100), all nickel-cadmium cells have an output voltage around 1.25 volts. This is the lowest cell output voltage of any commercially produced electrochemical couple. It takes more nickel-cadmium cells to produce a battery of any given voltage than any other chemical type of cell.

For example, a 12-volt battery composed of ni-cads will require the series use of 10 nickel-cadmium cells (1.25 volts per cell times 10 cells equals 12.5 volts). The same 12-volt battery made up of lead-acid cells would only require six cells (2.2 volts per cell times 6 cells equals 12.6 volts). The lower operating voltages of the nickel-cadmium reaction are in part offset by the cell's very low internal resistance. Ni-cads are much more capable of maintaining their voltage while delivering high rates of discharge than are the lead-acid type.

Voltage as a Function of Rate of Discharge

The nickel-cadmium cell's output voltage is dependent primarily on the rate of discharge of the cell. The faster energy is withdrawn from the cell the lower its voltage is. Temperature, state of charge, and type of cell construction are also minor factors affecting the output voltage of the cell.

Figure 3-1 gives the relationship of cell voltage to the rate of discharge for vented pocket plate ni-cads. Figure 3-2 gives the relationship of cell voltage to the rate of discharge for sealed sintered plate ni-cads. In both cases the temperature is assumed to be 78° F.

These graphs demonstrate the relatively flat output voltage curve that is characteristic of the nickel-cadmium cell, especially the sintered plate types. The voltage remains fairly constant until the cell is emptied, when it drops off sharply. The graphs also show the output voltage depression with higher rates of discharge.

The constant output voltage of ni-cads has one disadvantage to the users of small portable appliances such as flashlights. Ni-cads give very little warning when they are about to run out. In normal zinc-carbon dry cells, the voltage drops as the cell empties, causing the light to dim. The dimmer the light the less energy remains in the zinc-carbon cell. Ni-cads, however, keep about the same voltage until they are totally empty. In the ni-cad powered flashlight the bulb does not gradually dim, but goes out quickly. One cannot tell how much energy remains in the battery by its ability to power the load.

The output voltage, while under load, at which the cell is considered to be empty is called the *discharge cutoff voltage*. In actual practice most ni-cads are considered empty when between 10 and 20 percent of their capacity remains. Discharge beyond this point lowers the battery's efficiency and shortens its life. The discharge cutoff voltage is a

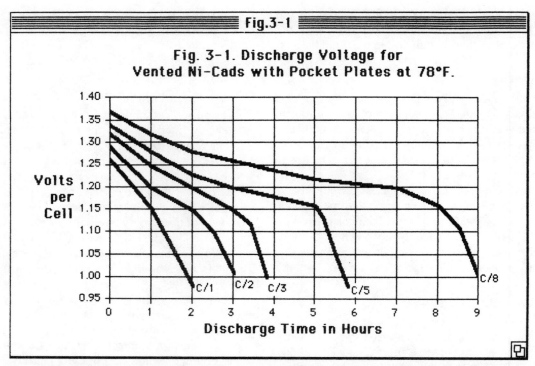

Fig. 3-1. Discharge Voltage for Vented Ni-Cads with Pocket Plates at 78°F.

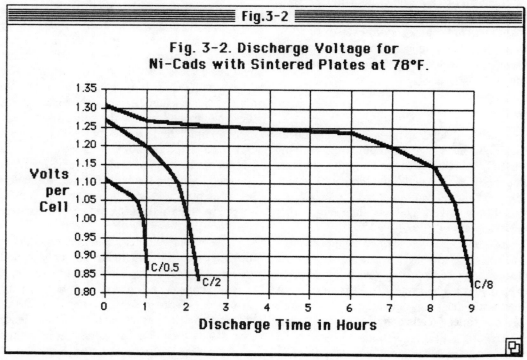

Fig. 3-2. Discharge Voltage for Ni-Cads with Sintered Plates at 78°F.

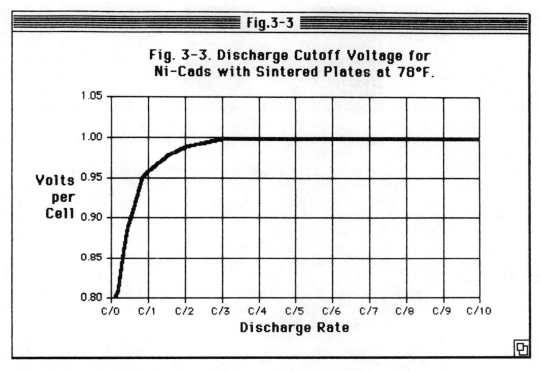

Fig. 3-3. Discharge Cutoff Voltage for Ni-Cads with Sintered Plates at 78°F.

function of the rate of discharge. Figure 3-3 gives this information for sintered plate ni-cads.

In most sintered plate ni-cads, the discharge cutoff voltage is specified at 1.0 volt for a discharge rate of C/5. For the pocket plate cell, the discharge cutoff voltage is 1.1 volts at the same rate. The cells can be cycled to even lower voltage levels if the discharge rates are slow, C/20 or less.

Voltage as a Function of Temperature

The voltage of a nickel-cadmium cell under discharge is effected by its temperature. The nickel-cadmium electrochemical reaction is effected by temperature in a different manner than the lead-acid reaction. Ni-cads operate much more efficiently at low temperatures than the lead-acid cells. The ni-cad maintains a more constant output voltage over a wider range of operating temperatures.

Figure 3-4 describes the relationship between the average output voltage of a ni-cad and its temperature. This graph gives the information for sintered plates and pockets plates (vented and sealed. This figure shows that all ni-cads have a range of optimum operating temperatures. If this range is exceeded the operating voltage drops off. This is true of ni-cads that are either too hot or too cold. Pocket plate cells demonstrate less voltage loss at high temperatures than do the sintered plate cells.

The output voltage of a ni-cad cell is highest when the cell temperature is between 60°F. and 90°F. This should be considered as the best range of temperature for ni-cad use. The ni-cad can be operated at much lower temperatures, but with reduced efficiency and voltage.

The output voltage of the sealed type of ni-cad decreases when the cell temperature is raised beyond 90°F. The vented pocket plate cell does not exhibit this voltage reduction over 90°F. Figure 3-4 also shows the decrease in output voltage that occurs with increasing the rate of discharge.

The "Memory Effect" of Ni-Cads

If a sintered plate nickel-cadmium cell is

repeatedly cycled shallowly it appears to lose its unused capacity. For example, if a sintered plate cell is repeatedly cycled to a 20 percent depth of discharge, the additional 80 percent of its capacity will become unavailable to the user. The battery will become "lazy" if its entire capacity is not used in each cycle. The amount of capacity lost to shallow cycling of the sintered plate cell depends on the number of repetitive shallow cycles it undergoes. The more shallow cycles the cell undergoes the larger the apparent loss of capacity.

This *memory effect* occurs only in the sintered plate nickel-cadmium cell. Pocket plate cells retain their complete capacity regardless of the depth of discharge to which they are subjected. The memory effect does not seem to occur in sintered plates which are randomly cycled; only in ones which are repeatedly shallow cycled.

An example of memory effect is the rechargeable electric razor's battery. The battery in the razor is continually being charged unless it is being used. A single shave will withdraw only about 25 percent of the available capacity of the battery. The razor is then returned to the charger where the battery is filled. After 20 or 30 of such shallow cycles, the razor's battery will deliver only enough energy for one shave, not the four or five shaves that were originally contained within its battery. The user of the razor might consider using four shaves from the razor before returning it to its charger for recharging.

Fortunately, the memory effect is not a permanent condition. A sintered plate cell can have this memory effect erased by several repeated complete discharge/charge cycles. This "exercising" of the cell should return the cell to its original capacity.

Rest Voltage as a Function of State of Charge

It is very difficult to determine the state of charge of a nickel-cadmium cell by measuring its voltage. This handy method of determining the state of charge of the battery is possible with lead-acid cells, but denied users of ni-cad cells.

The relatively flat output voltage curve of the ni-cads makes voltage measurement unreliable in

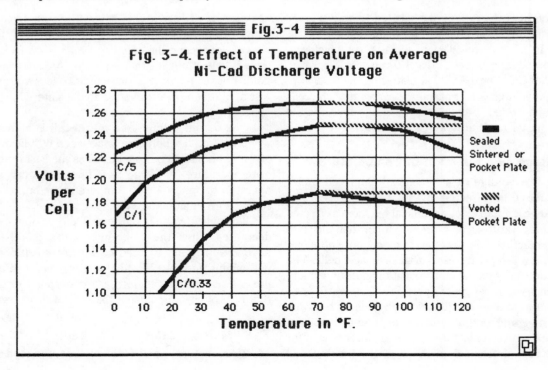

Fig. 3-4. Effect of Temperature on Average Ni-Cad Discharge Voltage

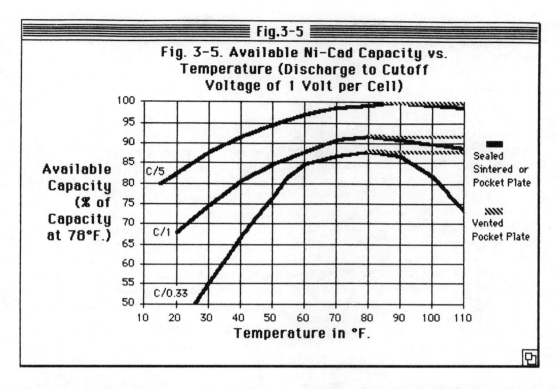

Fig. 3-5. Available Ni-Cad Capacity vs. Temperature (Discharge to Cutoff Voltage of 1 Volt per Cell)

determining their state of charge. The entire operating range of the output voltage is within 0.25 volts. Measurement inaccuracies and temperature can account for voltage differences greater than 0.2 volts.

Ni-Cad Battery Capacity as a Function of Temperature

The two factors determining the available capacity of a nickel-cadmium cell are the rate of discharge and temperature. Figure 3-5 gives the relationship between capacity and temperature for various rates of discharge. Discharge cutoff voltage is assumed to be 1.0 volt per cell.

In comparison to the lead-acid cell, the nickel-cadmium cell really shines at the lower temperatures. At 32°F. The ni-cad cell can deliver 90 percent of its rated capacity, while the lead-acid cell is capable of delivering only 70 percent its capacity. At −20°F. The difference is even greater; the ni-cad delivers 65 percent of its stored energy versus only 20 percent for the lead-acid cell.

Obviously those planning to use batteries for energy storage in very cold climates should consider the use of nickel-cadmium cells. The nickel-cadmium reaction has the widest range of operating temperatures of any type of commercially produced secondary cell.

The self-discharge rate of the nickel-cadmium cell is also affected by temperature. Ni-cads generally have very low rates of self-discharge in comparison with lead-acid cells. Ni-cads maintain this low rate over their entire lifetime. This is not the case with lead-acid cells, whose self-discharge rate increases radically with age.

Figure 3-6 shows the loss of capacity in ni-cads due to self-discharging as a function of temperature. This graph assumes that the battery is initially fully charged.

The self-discharge rate of the vented pocket plate ni-cad tends to be fairly constant over very long periods of time. The vented pocket plate ni-cad, stored at 72°F., will retain about 80 percent of its capacity after having been stored for two years. In the world of energy storage in chemical reactions, this is phenomenal. If the same cell is stored at 120° F. the self-discharge rate greatly increases. At 120° F. the cell would lose one-half its energy in less than 4 months. If the vented pocket plate ni-cad is used for long term storage then the temperature must be less than 80° F. Heat greatly accelerates the process of self-discharge.

The self-discharge rate of sealed sintered plate ni-cads is over 20 times greater than that of the vented pocket plate type. A sealed sintered plate cell will lose one-half its capacity in about two months at 72° F. If the temperature is above 110° F., this rate more than triples. This increased rate of self-discharge in sintered plate cells is due to their mechanical structure. The same factors which give the cell such a low internal resistance also make it discharge itself faster. Sealed sintered plate ni-cads used in portable appliances should be stored at room temperatures or less if the batteries are expected to hold their charge.

Ni-Cad Battery Longevity as a Function of Temperature

Operation of ni-cad cells at temperatures constantly above 100° F. will shorten the battery's life. Figure 3-7 demonstrates the relationship between battery life and temperature. This graph uses the vented pocket plate cell as a basis for information.

The effect of elevated temperatures on battery longevity is much less severe for the nickel-cadmium reaction than it is for the lead-acid reaction. Although high temperatures will shorten the life of most chemical batteries, the ni-cad is the least effected commercially produced secondary cell.

The sealed sintered plate ni-cad's longevity varies widely with temperature. The effect of temperature on these cells is primarily a function of the cell's internal construction. In general, the sealed sintered plate ni-cad is more sensitive to high temperatures than the pocket plate type.

CHARGING THE NI-CAD CELL

The nickel-cadmium cell is remarkably forgiv-

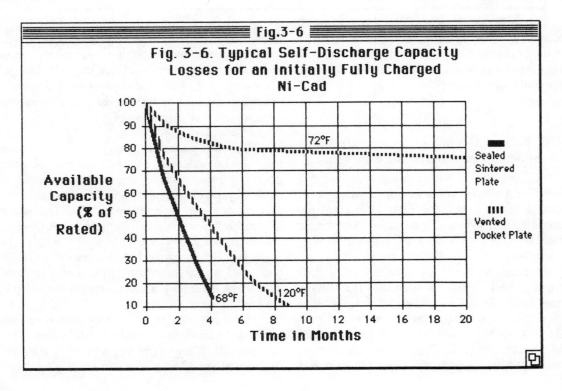

Fig. 3-6. Typical Self-Discharge Capacity Losses for an Initially Fully Charged Ni-Cad

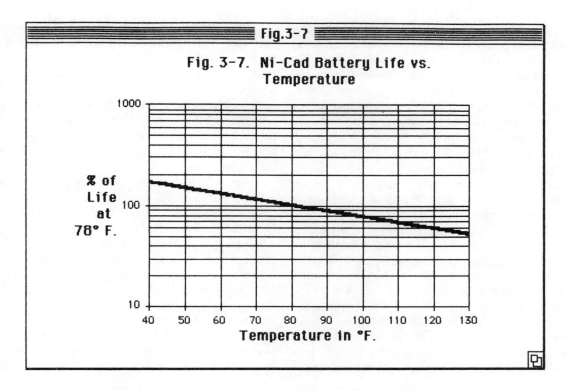

Fig. 3-7. Ni-Cad Battery Life vs. Temperature

ing at most temperatures and rates of discharge. One can push the ni-cad to the limits of its operating parameters without damage, with one notable exception—charging. Improper charging is the number one killer of nickel-cadmium cells.

It is important that all types of ni-cads be refilled as soon as they are empty. Ni-cads left in a discharged state for extended periods of time will become inefficient and short lived.

Rate of Charge

The rate at which the cell is recharged is a critical factor in the cell's longevity and efficiency. The ni-cad can be recharged at very high rates; some are able to be refilled in one hour. During rapid recharging care must be taken not to overfill the battery. All charge parameters must be compensated for the effects of temperature. The use of lower rates of charge (C/20 or less) does not require such precautions.

Nickel-cadmium cells are charged using current limitation as the primary method of charge control. Voltage control of the charge process is not successful due to the flat charge voltage profile offered by ni-cads. Temperature has more effect on the cell's voltage than does the state of charge.

Commonly used rates of charge for sealed sintered plate ni-cads are between C/4 and C/20. Rates under C/20 are not generally used. A good general rapid charge rate for the sealed sintered plate is the C/10 rate for 12 to 15 hours. A good rate for charging over extended periods (24 to 72 hours) is the C/20. At room temperatures, the sealed sintered plate ni-cad can withstand overcharging as long as the rate is C/20 or less.

Figure 3-8 shows the relationship between the rate of charge and the cell voltage for the sealed sintered plate ni-cad. These curves are very nonlinear. The voltage remains fairly constant over the entire charging process until the cell is almost full, when it rises slightly. When the cell is completely full the voltage begins to decrease. The information in the graph assumes the cell has a temperature of 78° F.

The sealed sintered plate ni-cad has a very low internal resistance. Assume the ni-cad is being charged by a constant voltage source. When the battery is full, most of the charging energy is converted to heat by the cell. This increase in heat within the cell lowers the internal resistance even further, which causes the battery to accept even more current, which causes even more heating, which again lowers the internal resistance, etc. What we have is a classic case of *thermal runaway*. The cells will overheat and gas. The cells will be ruined. Constant voltage charging of sealed ni-cads is not recommended.

Figure 3-9 demonstrates the effect of temperature on the maximum permissible overcharge rate in sealed nickel-cadmium cells. The higher the temperature the less the sealed cell will stand overcharging. If the sealed cell is to be left under continual charge, the higher the temperature is, the slower the rate of charge must be. A good general rate of charge for the sealed cell is C/20 at room temperatures. At this rate the cells may be left under charge for extended periods of time.

Pocket plate ni-cads are very resistant to damage from overcharging. They can withstand very high rates of charge. They are commonly recharged in a single hour (C/1). At this rapid rate care must be taken to remove the battery after one hour. Leaving the battery under charge at the C/1 rate for three or four hours will ruin it. A good safe rate for pocket plate ni-cads is the C/10 rate. If they are at room temperature, they can be charged almost indefinitely at the C/20 rate.

Figure 3-10 gives the relationship between the rate of charge and cell voltage in the vented pocket plate nickel-cadmium cell. The cell is assumed to be at 78°F. Notice that the voltage does not drop when the cell is full, as it does in the case of the sintered plate ni-cad. Vented pocket plate ni-cads do not participate in thermal runaway due to their higher internal resistance.

The rate at which a nickel-cadmium cell is charged has an effect on the efficiency of the cell's operation. Figure 3-11 gives the relationship between the rate of charge of a cell and the efficiency of the charge portion of the cell's cycle. The

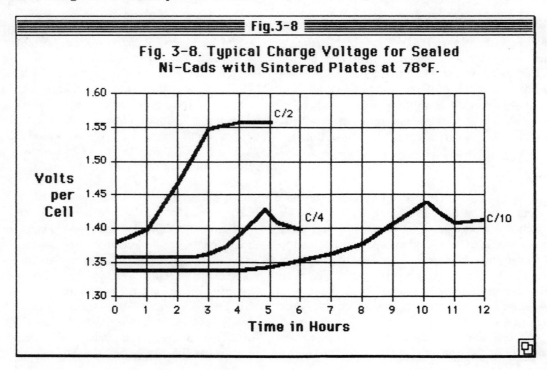

Fig. 3-8. Typical Charge Voltage for Sealed Ni-Cads with Sintered Plates at 78°F.

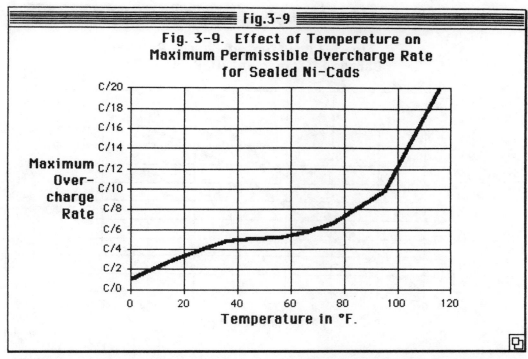

Fig. 3-9. Effect of Temperature on Maximum Permissible Overcharge Rate for Sealed Ni-Cads

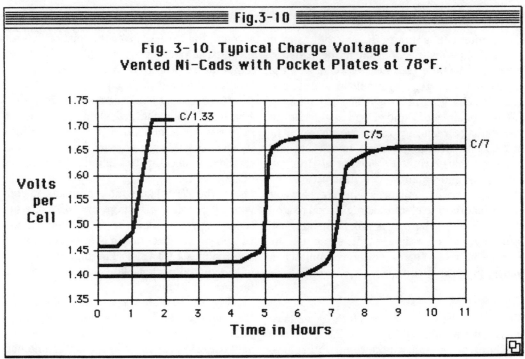

Fig. 3-10. Typical Charge Voltage for Vented Ni-Cads with Pocket Plates at 78°F.

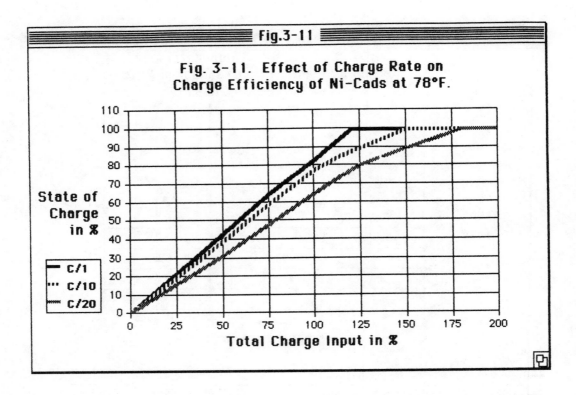

Fig. 3-11. Effect of Charge Rate on Charge Efficiency of Ni-Cads at 78°F.

nickel-cadmium cell is more efficient the faster it is charged. This is just the opposite of the lead-acid cell. Care must be taken because overcharging the ni-cad cell at high rates can result in violent cell damage. The charging process is one of balancing the higher efficiency of rapid charging against the possible damage resulting from overcharging at high rates. The efficiency picture is not that bad for the lower charge rates to warrant unattended rapid charging and potential cell destruction.

Manufacturers of appliances which recharge at rapid rates incorporate electronic regulation to stop the process when the battery is full. Some of these regulators operate on a time factor; some use the cell's temperature as a basis for determining when the charge process is complete.

Charge Voltage

While the cell is under charge, voltage measurement can be used to determine if the cell is fully charged. One must take temperature into account. Figure 3-12 gives the relationship between the charge voltage of a sealed ni-cad cell and temperature. Figure 3-13 gives the same information for a vented pocket plate ni-cad.

If voltage is to be used as an indicator of when a battery is fully charged, then the voltmeter used in the measurement must be very accurate. Differences of 0.02 volts are significant. The voltmeter should have accurate resolution in the neighborhood of 0.01 volts.

At the C/10 charge rate, the sealed ni-cad is full when it reaches a voltage of 1.45 volts at room temperature. Under the C/10 charge rate, the vented pocket plate ni-cad is full when the cell voltage reaches 1.60 volts. The higher finish voltage of the vented pocket plate cell is due to its higher internal resistance.

Figure 3-14 shows the relationship between float voltage and water consumption in vented pocket plate nickel-cadmium cells. As with lead-acid batteries, the most significant maintenance re-

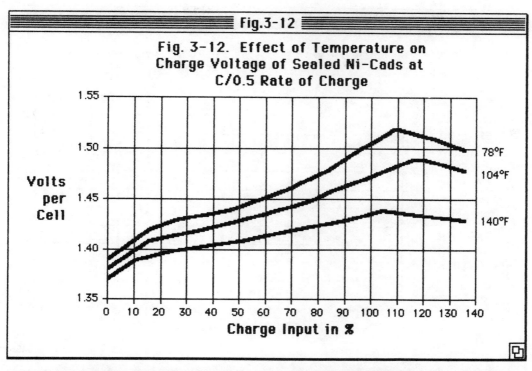

Fig. 3-12. Effect of Temperature on Charge Voltage of Sealed Ni-Cads at C/0.5 Rate of Charge

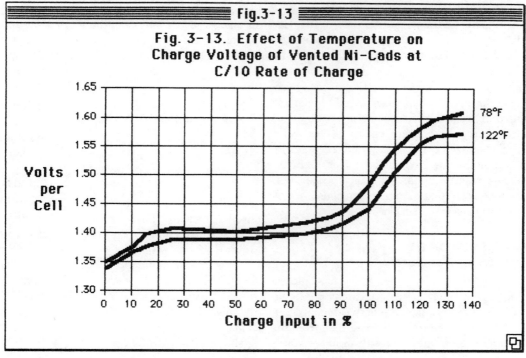

Fig. 3-13. Effect of Temperature on Charge Voltage of Vented Ni-Cads at C/10 Rate of Charge

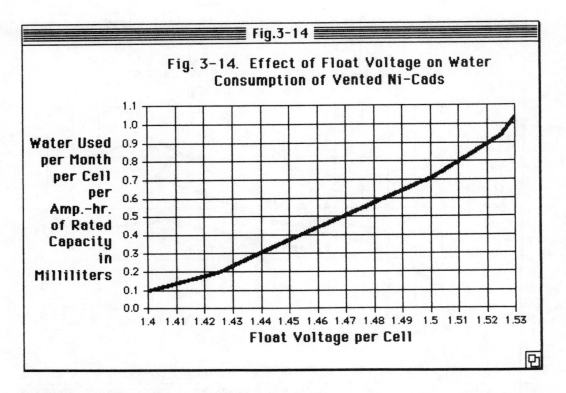

Fig. 3-14. Effect of Float Voltage on Water Consumption of Vented Ni-Cads

quirement is the addition of water to the cells. Use only distilled water to replace lost electrolyte. Be very careful not to introduce contaminants in the cells.

Water consumption is a function of many factors. Temperature, rate of charge, and type of cell construction are some factors. Figure 3-14 shows water consumption versus charge voltage for vented ni-cads which are voltage floated in standby service. In float service, the battery is not in use but held in reserve under constant charge. This float service results in higher water consumption than any other type of service.

Charge Efficiency as a Function of Temperature

The higher the temperature of a nickel-cadmium cell the more difficult it is to charge. Figure 3-15 gives the relationship between charge efficiency and temperature. As can be seen from the graph, higher temperatures result in lower states of charge for an equivalent amount of energy input to the cell. In other words, high temperatures result in lower efficiency. At temperatures over 110° F. it may become impossible to fully charge the cell.

Rejuvenation of Tired Ni-Cads

Over a period of time some ni-cads begin to lose some or all of their capacity. They become difficult to charge and will not hold a charge for long. The causes of this are mostly improper charging and allowing the battery to languish about in the discharged state. There are special charging techniques which can restore "tired" ni-cads to their original state.

The most common form of ni-cad failure is due to memory effect. This was detailed earlier. The same charge scheme of repeated deep cycling of the ni-cad works on cells which show reduced capacity due to age or undercharging. The process is simple. Fully discharge the cell at the C/2 rate or faster. Once the cell is fully discharged, fill it at as rapid a rate as is prudent. Be careful not to overcharge the cell at high rates. Repeat this process up to five times or until the ni-cad regains its full capacity. Some cells will not respond to this treatment and will require more violent measures to restore them to service.

If the ni-cad does not respond to repeated rapid deep cycling then it may be suffering from a condition known as *dendritic growth*. This condition is fully described in the technical section at the end of this chapter. In short, the active materials in the cell (nickel and cadmium) form metallic whiskers known as *dendrites* which can internally short out the cell. This condition can be recognized by the fact that the cell behaves as a dead short. It will show little (less than 0.8 volts) or not output voltage after charging. Charging, even at very rapid rates, will not raise the cell's voltage much above 1 volt while under charge. In light cases of dendritic growth the cell may hold some energy but it will self-discharge rapidly in one or two days. The only cure for dendritic growth is a procedure known as *zapping*.

When a ni-cad is zapped it has a very large current run through it for a very short period of time. Currents as high as C/0.02 for a period of less than 1 second can be used. Zapping is a dangerous procedure. Great care must be taken to assure safety. If the rate of electron flow through the cell is too high or maintained for too long, cell damage will certainly result. The cell will vent and be ruined. It may even explode. Always wear gloves and eye protection when zapping ni-cads.

Here is an example of zapping a sealed D-sized ni-cad. The cell should be placed in a refrigerator and cooled to about the freezing point of water. The cell is cooled so that the possibility of heat damage during zapping is minimized. Use a fully charged 12-volt lead-acid battery of at least 100 ampere-hours capacity as a zapping power source. Attach a wire to the positive terminal of the lead-acid battery. This wire should be of 12-gauge size or larger. Place the negative pole of the ni-cad firmly on the

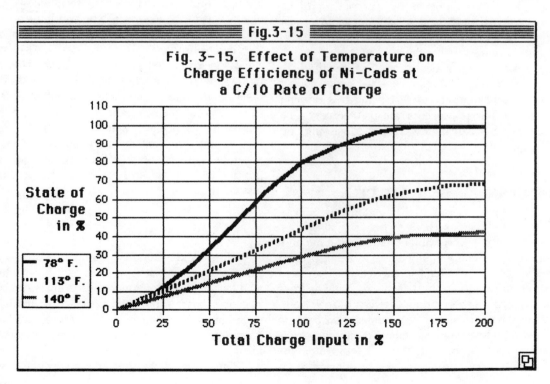

Fig. 3-15. Effect of Temperature on Charge Efficiency of Ni-Cads at a C/10 Rate of Charge

negative pole of the lead-acid battery. Briefly, very briefly, touch the wire connected to the positive pole of the lead-acid battery to the positive pole of the ni-cad. There will be much sparking. Time of contact should not exceed 1 second. In fact it should be as short as possible.

Zapping destroys the dendrites shorting out the cell. For this brief time, less than one second, currents in excess of 100 amperes will flow through the ni-cad cell. These high currents will vaporize the dendrites within the ni-cad and hopefully not destroy the cell.

After the zapping procedure, the cell will become very warm. Allow the cell to rest and cool for at least 24 hours before repeating the zapping procedure. Most dendritic cells will respond in less than three zappings. For zapping cells of C size the procedure is the same only use a six-volt lead-acid battery as a power source. For zapping AA cells use a two-volt lead-acid battery. Most power supplies, with the exception of the lead-acid battery, will not deliver enough current to vaporize the dendrites. If you are going to use a power supply be sure it can deliver at least 200 amperes into a short circuit. Be sure the supply has a direct current (dc) output; ac output will ruin the cell.

Personal experimentation has resulted in salvaging 67 out of 70 D- and C-sized sealed pocket plate nickel-cadmium cells by using this zapping procedure. This is a success rate of 90 percent. The three cells not salvaged were lost to gassing and cell rupture.

DISCHARGING THE NI-CAD CELL

The low internal resistance of the nickel-cadmium cell gives them the ability to be discharged very rapidly. In fact, if you examine a ni-cad flashlight cell you will see a warning printed on the case. If a ni-cad is carried in a person's pocket and a metal object such as coins or keys were to make a short circuit between the cell's output terminals, then the metal object can become very hot and cause burns. Short circuited ni-cads can become discharged in much less than one minute.

Discharge Rates

Small ni-cad batteries often are discharged at rates which empty them in less than 30 minutes. Unlike the lead-acid cell, this does not damage them. Discharge rate of between C/3 and C/20 are common in sintered plate ni-cads within portable appliances.

The vented pocket plate ni-cad, being a large storage cell, is discharged much more slowly. This is more a reflection of its application than its capabilities. These cells are discharged at rates between C/10 and C/600. In storage applications, especially those fed on alternative energy, the battery is sized with enough capacity to last at least several days and in some cases several months. With such large capacities being used for storage, most average discharge rates are quite slow.

Self-Discharge

In large storage ni-cads, the self-discharge rate is not as limiting a factor in long term energy storage as it is in the lead-acid type. With a lead-acid battery, it is not efficient to store any more energy than can be used in two or three weeks. This is not the case with vented pocket plate nickel-cadmium batteries. The ni-cad discharges itself at less than one-third the rate of a new lead-acid cell. If the lead-acid cell is old, then the ratio of self-discharge rates is more like 1/10. The constantly low self-discharge rate in these ni-cads makes it possible to store several months of energy efficiently. The only real limiting factor is the size of the bank account paying for the cells.

Sizing Ni-Cad Batteries

With nickel-cadmium batteries, having too much capacity is not a problem as it is with lead-acid batteries. In general, the problem with sizing ni-cads is having too little capacity. Large nickel-cadmium storage cells are so expensive that most users tend not to buy enough capacity to meet the need.

The source of charging energy must be considered as well as the demand for energy. Systems which depend on solar or wind power sources must be oversized in capacity to allow for cloudy or windless periods. As with lead-acid cells used in alternative energy service, the use of a motorized power source should be considered as a backup to the alternative energy sources. In many cases, the added cost of a motorized backup charger is less than the cost of oversizing the nickel-cadmium battery pack.

A balance can be reached between the different factors of demand, battery cost, and self-discharge. Demand can be estimated by using the information in Chapter 9. Generally, ni-cad batteries are sized to last the user between one and six weeks before recharging. Packs which have less than a one-week capacity usually spend most of their time discharged in alternative energy applications. This results in reduced battery life.

TECHNICAL DATA FOR NI-CAD CELLS

The limits of a nickel-cadmium battery's performance are determined by the electrochemistry of the cell. An understanding of the basic chemistry of the cell will help us to use these batteries more efficiently.

Chemical Composition

Each nickel-cadmium battery has two output terminals, or *poles*. The positive pole is known as the anode. The negative pole is known as the cathode.

The anode of a nickel-cadmium cell is composed of two nickel compounds. These nickel compounds gradually change as the battery is charged or discharged. When the cell is fully charged the nickel compound existing on the anode is nickel oxide hydroxide, NiO(OH). When the cell is fully discharged the anode becomes nickel hydroxide, $Ni(OH)_2$. A cell at a 50 percent state of charge would have each of these nickel compounds present in equal quantities.

The cathode of the nickel-cadmium cell consists of the metallic element cadmium (Cd) when the cell is fully charged. When the cell is fully discharged, the cathode consists of cadmium hydroxide [$Cd(OH)_2$].

The electrolyte in the nickel-cadmium reaction is a 25 percent to 35 percent solution of potassium hydroxide [K(OH)] in water. This electrolyte does not enter into chemical change with the active materials of the cell. It exists as a catalyst, merely providing a medium for ion exchange within the cell. As such, the specific gravity of the electrolyte in a ni-cad cell does not change with the state of charge of the cell.

The relatively high costs of cadmium and nickel are, in part, the cause of the ni-cad's higher price tag. Cadmium and cadmium vapors are toxic, adding to the manufacturing complexity of the ni-cad.

Discharge Reactions

As a nickel-cadmium cell is discharged, the chemical composition of both the anode and the cathode changes. The electrons are transferred from the cathode to the anode via the external load.

During discharge, the anode gradually changes from nickel oxide hydroxide [NiO(OH)] to nickel hydroxide [$Ni(OH)_2$] due to the addition of the electrons from the cathode. The cathode, which is losing electrons, changes from cadmium (Cd) to cadmium hydroxide [$Cd(OH)_2$].

DISCHARGE

Anode: $2NiO(OH) + 2H^+ + 2e^- \rightarrow 2Ni(OH)_2$
Cathode: $Cd + 2(OH)^- - 2e^- \rightarrow Cd(OH)_2$

The $2H^+$ and the $(OH)^-$ radicals are ions from the electrolyte. They are available from the hydrolysis of the electrolyte. The electron transfer through the electrolyte completes the circuit within the cell.

Each cadmium atom which reacts chemically by changing into cadmium hydroxide [$Cd(OH)_2$] releases two free electrons. This is the source of the ni-cad's power. These two electrons flow through the external load and arrive at the anode where they deoxidize the nickel oxide hydroxide [NiO(OH)] into nickel hydroxide [Ni(OH)$_2$]. This reaction is a demonstration of the tendency of all materials to seek more stable electronic configurations.

The nickel-cadmium reaction is an oxidation-reduction or *redox* reaction. During discharge, reduction occurs at the anode and oxidation at the cathode. This type of redox reaction is characteristic of all electrochemical cells.

Charge Reactions

The charging process in the nickel-cadmium cell is simply the reverse of the discharge process. Voltage is applied to the cell's terminals and the flow of electrons is reversed within the cell. The compounds forming the anode and cathode return to their original fully charged states.

During the charging process, the anode changes from nickel hydroxide to nickel oxide hydroxide. The cathode changes from cadmium hydroxide to cadmium. The anode is oxidized and the cathode is reduced. When these chemical changes are complete, i.e. both the anode and cathode have returned chemically to their original charged states, the cell is considered full.

CHARGE REACTIONS
Anode: $2Ni(OH)_2 - 2e^- \rightarrow 2NiO(OH) + 2H^+$
Cathode: $Cd(OH)_2 + 2e^- \rightarrow Cd + 2(OH)^-$

These equations reflect the direction of electron flow within the cell while it is under charge. The anode is depleted of electrons while the cathode is being filled chemically with an excess of electrons.

Note the hydrogen evolving at the anode and the hydroxide radical forming at the cathode. Under normal charging conditions, these two chemicals will combine to form water. If the cell is charged too rapidly for all the hydrogen and hydroxide molecules to form water, then some may escape the cell as gases. Hydrogen gas is highly explosive. Care must be taken not to charge ni-cads too rapidly or they will gas. In the case of a sealed cell, the evolution of these gases will increase the internal pressure within the cell and cause it to vent. The cell is ruined. Vented cells can survive gassing; however, these gases can be explosive. Users of vented ni-cads should always provide adaquate ventilation for the battery pack. Overcharging and too rapid charging of vented cells causes greatly increased water consumption.

Figure 3-16 gives a schematic representation of the chemical changes which occur during cycling the nickel-cadmium cell. This figure is not an actual physical representation of the mechanical construction of a ni-cad cell.

Cell Contamination and Water Loss

The only routine maintenance to be performed on vented nickel-cadmium cells is the addition of water. This is done to replace water lost from the electrolyte due to hydrolysis. One might suppose this to be a simple chore to perform well. However, many expensive ni-cads are ruined by contamination during this maintenance.

All batteries are chemical machines. The inside of a cell must be kept chemically pure. Pure distilled water is the only type of water that should ever be added to the cells. Tap water may contain metallic salts and other dissolved impurities which can ruin the inside of a ni-cad cell in just a few months.

Before water is added to the cells the tops of the batteries must be very clean. Once the vents are opened for filling, small particles of dust and dirt may find their way into the cell. All equipment used to handle the distilled water must be very clean. If the funnel used to fill the cells is dusty then the cells will be contaminated. Care must be taken that the vent caps, once removed, are not placed on dusty surfaces.

When you open your cells for water addition, envision yourself to be a surgeon. In surgery a doctor takes extreme care not to contaminate or infect the patient. All instruments and operating surfaces are antiseptically clean. If your batteries are to deliver their full service, their insides must be just as clean as a surgeon's operating room.

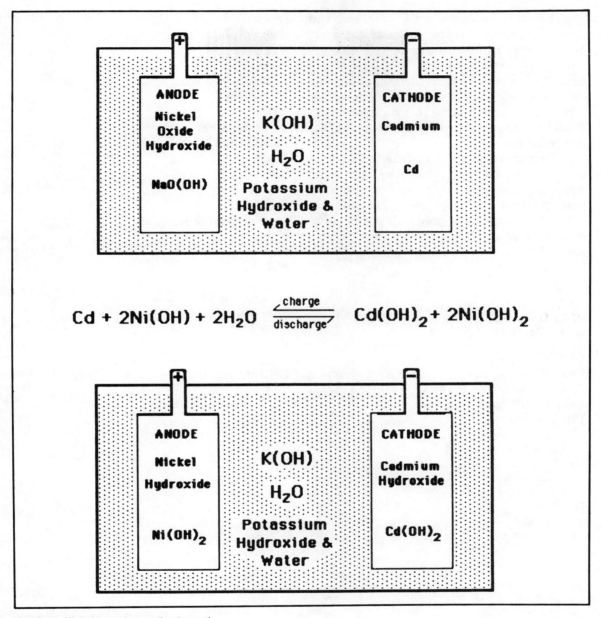

Fig. 3-16. Nickel-cadmium cell schematic.

Dendritic Growth

Every time a cell is cycled the chemical composition of both its anode and cathode changes. After hundreds of cycles the materials on the plates gradually become less regular in physical form. The fields surrounding these atoms during chemical change favor the linear physical form. The plates start to grow very thin *whiskers*. These thin threads of nickel and cadmium are known as *dendrites*.

There is no problem with these dendrites until they make a direct contact between the anode plates and the cathode plates within the cell. As a direct short circuit within the cell, the dendrites effectively ruin the cell. Cells which have dendritic shorts are electrically low resistance resistors. They will not store energy as the path through the dendrites has much less resistance than the path of normal cell operation.

The method for removing these dendrites was detailed earlier. This procedure essentially vaporizes the dendrites by running very high currents of electricity through them. The danger involved in this process is gassing. If the procedure is applied properly, the chance of gassing is minimal.

NI-CAD BATTERY MANUFACTURERS

SAB NIFE, Inc.
Post Office Box 100
George Washington Highway
Lincoln, RI 02865
401-333-1170

Panasonic Industrial Company
Division of Matsushita Electric
Corporation of America
Post Office Box 1511
Secaucus, NJ 07094
201-348-5266

Varta Batteries, Inc.
150 Clearbrook Road
Elmsford, NY 10523
914-592-2500

Gould, Inc.
Portable Battery Division
931 North Vandailia Street
St. Paul, MN 55114
612-645-8531

General Electric Company
Post Office Box 861
Gainsville, FL 32602
904-462-3911

Chloride, Inc.
Mallard Lane
North Haven, CT 06473
203-777-0037

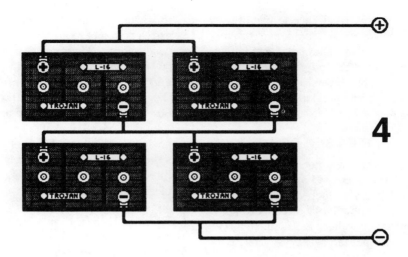

Edison Cells

The nickel-iron cell has been in limited commercial production for over 80 years. It is known as the *Edison cell* because Thomas A. Edison was the primary developer of the original nickel-iron battery. The Edison cell has developed a mystique surrounding it. For reasons unknown, users of battery stored energy seem to feel that the Edison cell is a superior type of cell, which has unfortunately fallen into disuse. The facts do not support this opinion, which is probably due to the "grass is greener" syndrome. The mystique of the Edison cell's superiority is fostered by its unavailability and its name.

The Edison cell has nickel and iron as its active materials. The electrolyte is a solution of potassium hydroxide in water. The operating voltage of an Edison cell is about 1.25 volts. Its nominal energy efficiency is around 60 percent less than either the nickel-cadmium or the lead-acid type. This lower efficiency is due to the Edison cell's relatively higher internal resistance.

The Edison cell is chemically very similar to the nickel-cadmium cell. Both are alkaline cells using nickel as one of the active materials. Due to the Edison cell's operating characteristics and expense, it has not been widely popular or manufactured in large numbers.

The Edison cell has several drawbacks which led to its disuse. The cells have very rapid rates of self-discharge, high expense, and a narrow temperature operating range. Nickel-cadmium and lead-acid type cells offer better performance at lower prices. These combined factors make the Edison cell a very rare item.

PHYSICAL CONSTRUCTION

The Edison cell is similar to the lead-acid cell and the vented nickel-cadmium cell in physical construction. The active materials are held on parallel plates within the cell. These plates are assembled into stacks within the battery case, much in the same manner as a car battery. The electrolyte is liquid and the cell is vented for pressure relief and water addition.

The cell case is now usually made from thermoplastic; however, glass and hardened rubber have been used in the past. They are rectangular in shape, with the terminals and vent on the top. The Edison cell is capable of vertical operation only. The cells are quite heavy in relation to their size. Each cell usually has its own case.

APPLICATIONS FOR EDISON CELLS

The Edison cell was very popular for motive power around 1910. It was used for engine starting and for powering early electric vehicles. The cell rapidly declined in popularity as the lead-acid type was developed. The Edison cell is much more expensive to use than the lead-acid type. Most jobs that were done in the past using Edison cells are now being done more cheaply using the lead-acid cell.

Edison cells are currently in use in electric mine vehicles. Some railroads use Edison cells for lighting and communication backup on trains. In general though, the use of Edison cells has almost ceased in the United States and is minimal in other parts of the world.

Edison cells are practical in applications that require a very high degree of mechanical ruggedness. They shine in high vibration environments like the caboose on a freight train. There is speculation that the Edison cell may be suitable to power consumer electric vehicles. Electric vehicle batteries are discharged and charged daily. In this type of service, the high rate of self-discharge of the Edison cell is not a disadvantage because the energy is held within the battery for only short periods of time. Westinghouse, Eagle-Picher, and the Swedish National Development Co. are involved in research and development of the Edison cell for electromotive service. These prototype cells are strictly experimental and are not currently commercially available. Considering the design requirements of vehicular service, it is unlikely that these cells will be applicable in alternative energy or home service.

EDISON CELL COST

At the current time there are no commercial manufacturers of Edison cells in the United States. The prices stated here are based on units made in Europe and Japan. These prices are FOB, considering the weight of these cells transportation could well add 25 to 40 percent to these prices.

The cost of an Edison cell battery is about $242 to $460 per kilowatt-hour. These prices reflect the relatively high cost of nickel. The lead-acid type of cell can do most jobs an Edison cell can do at a fraction of the price. It is highly unlikely that the Edison cell will return to the marketplace in any quantities. The nickel is more effectively used in the nickel-cadmium cell.

Edison cells have been manufactured in sizes between 6 ampere-hours and 1000 ampere-hours. The imported cells are available in a capacity of 250 ampere-hours. Once again, the actual effective capacity of any battery is greatly influenced by temperature and discharge rate.

LIFE EXPECTANCY

The Edison cell is a long lived type of rechargeable cell. It has a cycle lifetime of up to 2000 cycles. Its calendar lifetime is around 7 to 12 years. If the battery is strictly in float service it may last as long as 25 years. These figures are for batteries in optimum conditions. Temperature, cycle rates, and maintenance are big factors in any battery's longevity.

The Edison cell is very resistant to permanent damage from overcharging and overdischarging. As such, it has high survival rates in applications where the cells are not used carefully. Its longevity suits it for long term float type service.

EDISON CELL CHARACTERISTICS

The Edison cell displays the same general operating characteristics as the vented nickel-cadmium cell. The reactions are very similar chemically. The ni-cad has a wider range of operating temperatures and very much lower self-discharge rate. The Edison cell has an output voltage of 1.25 volts, the same as a ni-cad cell. This is reasonable since both types of cells are chemically based on the oxidation of nickel.

Voltage

The voltage at the terminals of an Edison cell while it is operating is a function of many factors. Temperature, rate of charge/discharge, state of charge, and cell condition are some factors affecting the cell's voltage.

Figure 4-1 presents the output voltage of an Edison cell in relation to discharge rate and state of charge. The cell temperature is assumed to be 78° F. The voltage of an Edison cell is between 1.4 to 1.5 volts when it is fully charged. This voltage gradually becomes less as the cell is discharged. At a state of charge of 50 percent the cell's voltage is about 1.2 volts. When the cell is fully discharged the voltage is less than 0.9 volts. These cells are usually considered empty when the output voltage is lower than 1 volt.

The nickel-iron cell has higher internal resistance than does the nickel-cadmium cell. This higher internal resistance gives the Edison cell more voltage fluctuation as it is discharged, and generally lower efficiency.

Temperature

The Edison cell is more effected by temperature than its close cousin, the ni-cad. This is in part caused by its higher internal resistance. Figure 4-2 gives the relationship between output voltage, temperature, and state of charge.

The Edison cell has effectively lost half its energy when the cell temperature is below 32° F. This characteristic of the Edison cell makes them poor choices for operation in cold locations. The optimum temperature range for Edison cell operation is between 50° F. and 90° F.

The cell is capable of operation at temperature's above 110° F. without damage. At high temperatures the Edison cell has very high self-discharge rates which adversely affect the efficiency of the cell.

CHARGING THE EDISON CELL

The Edison cell is charged in much the same fashion as other rechargeable cells. Voltage and

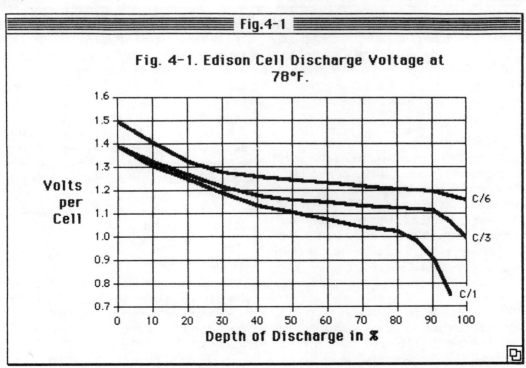

Fig. 4-1. Edison Cell Discharge Voltage at 78°F.

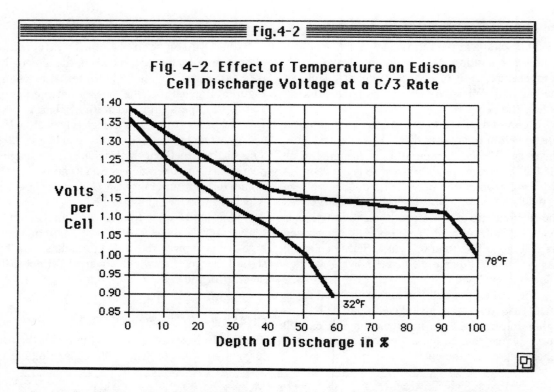

Fig. 4-2. Effect of Temperature on Edison Cell Discharge Voltage at a C/3 Rate

rate of charge are factors which affect the efficiency and longevity of the battery. One distinguishing factor of the Edison cell is the constant gassing which takes place during the entire charge cycle. This is unique among rechargeable batteries.

Rate of Charge

The Edison cell is capable of being charged at rapid rates. Rates of C/1 to C/10 are acceptable. If high rates of charge (over C/3) are used, care must be taken to keep the cells cool. The relatively high internal resistance of the Edison cell causes much of the charge energy to be dissipated in heat. Rapid rates of charge and high temperatures can cause severe gassing in the Edison cell.

Gassing

Gassing occurs throughout the entire charge cycle in the Edison cell. This is different from the lead-acid and the vented ni-cad cells, which gas only when overcharged or charged too rapidly. This constant gassing under charge makes the Edison cell a safety risk. Hydrogen and oxygen are emitted from the battery during the entire charge cycle. These gases are explosive. Care must be taken to supply adequate ventilation for the Edison cell.

Each ampere of current will generate 0.00027 cubic feet of hydrogen gas per minute, per cell. If the temperature is over 110° F., this rate of gassing is more than doubled. Edison cells are large consumers of distilled water. The energy loss in the hydrolysis of water is a primary area of energy inefficiency in the Edison cell.

Charge Voltage

The Edison cell demonstrates a relatively flat voltage curve while under charge. Figure 4-3 gives the relationship between cell voltage, rate of charge, and state of charge. The cell temperature is assumed to be 78° F. At the C/10 rate the Edison cell is considered full when the cell voltage reaches around 1.6 volts. The temperature is assumed to be 78° F.

At higher temperatures, the cell will be full at lower voltages. The voltage curve tends to flatten when the cell is almost full (over 80 percent). This makes it difficult to determine the exact state of charge using voltage as an indicator.

Those considering using a constant voltage to charge the Edison cell for extended periods of time (float service) should use a voltage of around 1.45 to 1.5 volts per cell. If the temperature is over 90° F., then this voltage should be reduced to 1.35 volts per cell. In general, the Edison cell excels in float service. It has high longevity and is relatively tolerant of overcharging. Cells in float service should have their water checked weekly. Edison cells always seem to be thirsty.

DISCHARGING THE EDISON CELL

The Edison cell cannot deliver energy as fast as either the nickel-cadmium cell or the lead-acid cell. It does best at slower rates of discharge. Since the rate of discharge is dependent on the batteries capacity, the battery pack must have more capacity for any given load.

Rates of Discharge

The Edison cell is capable of discharge rates between C/1 and C/10. Discharge rates faster than C/3 are inefficient due to low voltage in the last 25 percent of the discharge cycle.

The relatively high internal resistance of the Edison cell makes them unsuitable for discharge rates faster than C/0.5. Rapid discharge of these cells results in low operating voltage and reduced effective capacity.

Self-Discharge

The major drawback of the Edison cell is its very rapid rate of self-discharge. This high rate makes these cells unsuitable for longterm energy storage.

The Edison cell loses 10 percent of its rated

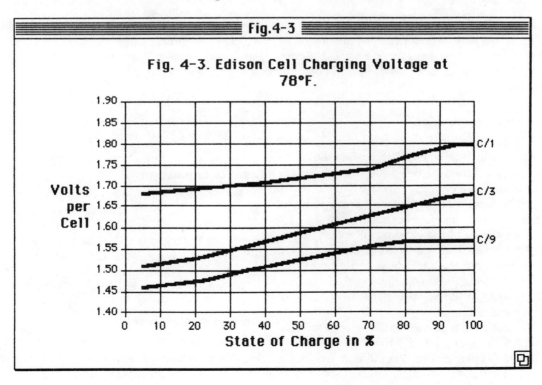

Fig. 4-3. Edison Cell Charging Voltage at 78°F.

capacity to self-discharge within one week. That is 40 percent of the stored energy within the cell during a month's time. The temperature is assumed to be 78° F.; if it is warmer, then the rate greatly increases. This is over twice the self-discharge rate of the lead-acid type and eight times that of the vented ni-cad. Edison cells should be completely cycled in several days time. If energy is stored within them for longer than a week, the efficiency picture becomes very grim. This type of battery is not suitable for deep cycle applications.

Nickel-iron cells are really not the type to use in alternative energy systems. Their high rate of self-discharge alone is enough to disqualify them. Add to this their expense and narrow temperature range and it is easy to see why they are rarely commercially produced.

TECHNICAL DATA FOR THE EDISON CELL

The chemistry of the Edison cell is similar to that of the nickel-cadmium cell. Consult Chapter 3 for additional technical details on alkaline cell reactions.

Figure 4-4 shows the chemical operation of the Edison cell in schematic form. This picture is not a physical representation of the mechanical interior of the cell, but a schematic of its chemical operation.

Chemical Composition

The anode of an Edison cell is composed of nickel oxide hydroxide [NiO(OH)] when the cell is fully charged. When the cell is fully discharged the anode is composed of nickel hydroxide [Ni(OH0$_2$]. These chemical compositions are the same found on the anode of the nickel-cadmium cell.

The cathode of the Edison cell is made up of iron (Fe) when the cell is fully charged. In a discharged state the cathode is ferric hydroxide [Fe(OH)$_2$]. Again, this is similar to the ni-cad cell, with the only difference being the substitution of iron for cadmium.

The electrolyte is a 20 to 25 percent solution of potassium hydroxide [K(OH)] in water. Small amounts (less than 2 percent) of lithium hydroxide [Li(OH)] are also added to the electrolyte. The electrolyte is merely a medium for electron transfer in the Edison cell. The electrolyte is a catalyst and does not participate in chemical change with the active materials. The specific gravity of the electrolyte of an Edison cell has no relation to the cell's state of charge.

Discharge Reactions

As in most electrochemical cells, the charge/discharge reactions in the Edison cell are oxidation-reduction (redox) reactions. The anode is reduced and the cathode is oxidized during the discharge portion of the cycle. The reaction is very similar to that of nickel-cadmium cell.

DISCHARGE REACTIONS

Anode: $2NiO(OH) + 2H^+ + 2e \rightarrow 2Ni(OH)_2$
Cathode: $Fe + 2(OH)^- - 2e^- \rightarrow Fe(OH)_2$

Charge Reactions

The charge reactions are the opposite of the discharge reactions. During the charge portion of the cycle the anode is oxidized and the cathode is reduced. The direction of electron flow through the battery is reversed. The active materials return to their fully charged states.

CHARGE REACTIONS

Anode: $2Ni(OH)_2 - 2e^- \rightarrow 2NiO(OH) + 2H^+$
Cathode: $Fe(OH)_2 + 2e^- \rightarrow Fe + 2(OH)^-$

Cell Contamination

The Edison cell requires frequent addition of water to the electrolyte to make up the water lost during charging. Edison cells should have their electrolyte level checked weekly. Use only distilled water and be sure not to introduce contaminants into the cells. Keep the tops of the cells clean, as well as equipment used to handle the electrolyte.

Electrolyte Replacement

The specific gravity of the electrolyte gradually decreases with age in the Edison cell. It is a com-

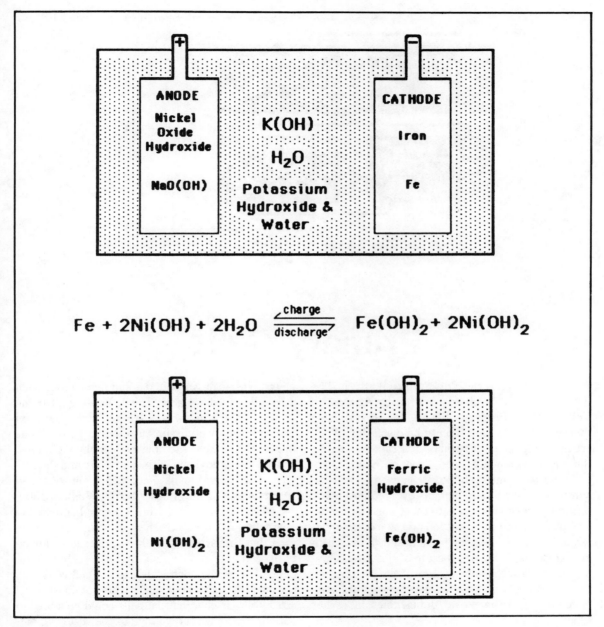

Fig. 4-4. Edison cell schematic.

mon practice to replace the electrolyte once or twice during the battery's lifetime. This electrolyte replacement is peculiar to the Edison cell, and is not performed on either the lead-acid or nickel-cadmium cells.

The replacement is accomplished by completely discharging the battery and shorting its output terminals. The old electrolyte is drained from the cells and a *renewal solution* is added. The battery is then charged at the C/10 rate for 24 hours.

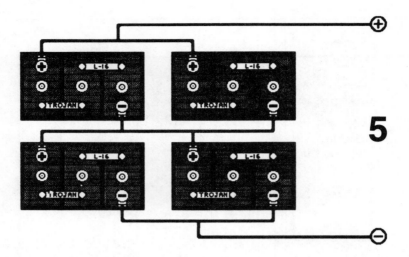

Primary Cells

The primary cell is the oldest type of battery. The first primary cell was constructed by Allesandro Volta in 1800. Primary cells are designed to have their energy used and then to be discarded. They are not designed to be recharged. Primary cells are the most commonly used type of battery, accounting for some 20 billion cells used worldwide annually. The primary cell has evolved into many different types and forms to meet specific needs. The physical and chemical construction of these different families of cells vary widely and sorting them out can be confusing. It is best to begin at the beginning and follow the development of the primary cell through time. Each type that has evolved has been developed in response to deficiencies in earlier types.

The battery constructed by Volta in 1800 was the first continuous source of electric current. Volta's primary battery was constructed of silver (cathode) and zinc (anode) disks separated by porous wafers of leather or paper saturated with electrolyte. Volta used a variety of chemicals for the electrolyte—salt water and alkaline solutions such as lye dissolved in water. The disks were piled one on top of the other with a leather wafer between each of the two dissimilar metallic disks. The piled disks, which resemble a stack of poker chips, was then wetted down with electrolyte and the battery was operational. Figure 5-1 illustrates a "Voltaic Pile" as first described by Volta in 1800.

Volta also noted the essential operating characteristics of electrochemical cells. He discovered that the larger the cell was, the more capacity it had. He also noted that the cell voltage depended on the chemicals used in cell construction and that cell voltage had nothing to do with cell size. Volta discovered that cells could be added in series to increase the voltage of the resulting battery. All in all, Volta is the father of the electrochemical battery.

Volta's battery was revolutionary; it fostered a number of major developments in physics and chemistry. The battery made possible experiments by Humphrey Davies in 1813, investigating the

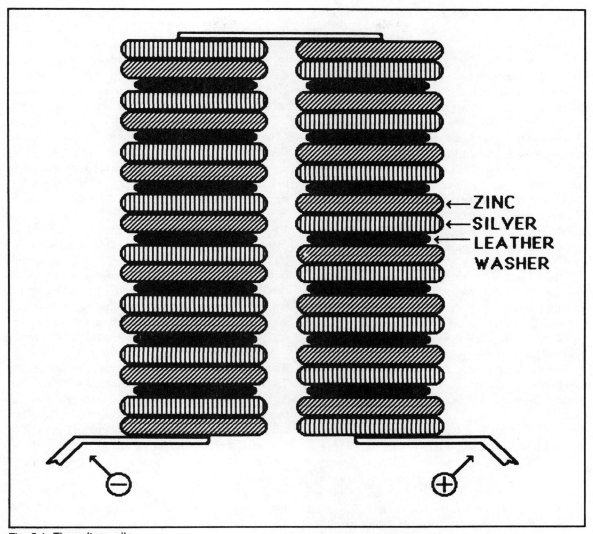

Fig. 5-1. The voltage pile.

nature of matter. The battery also made possible the experiments of Michael Faraday in 1831, resulting in breakthroughs in electricity and magnetism and the development of the generator. The battery during this period of time was used only by scientists and researchers. Batteries were thought of as novel devices having no real practical purpose. Battery design developed very slowly in this rarefied atmosphere.

The next major advance in battery technology was initiated by Georges Leclanche in the 1860s. Primary cells were being developed during this period with materials differing from the original voltaic pile. Batteries were being used in greater numbers and the push was on to produce cheaper and better cells. Obviously silver is inherently too expensive for mass production of large batteries. Leclanché developed a primary cell using a zinc rod for the anode and manganese dioxide mixed with powdered carbon for the cathode. The electrolyte was a solution of ammonium chloride dissolved in water. This primary wet cell was called the

"Leclanché" cell or the zinc-carbon cell. This cell is the historic forerunner of the modern ubiquitous zinc-carbon dry cell.

By the 1880s, the Leclanché cell was in common usage powering a variety of electrical devices. The Leclanché cell was essential in the development of both the telephone and telegraph. The Leclanché cell was also developed in a dry form. The cell was encased in a zinc cylinder and the electrolyte was reduced to a paste. This battery was the first easily portable power. These early *dry cells* were not very long lived and did not store very much power, but they ushered in a new age of portable electrical devices. The flashlight was developed around the turn of the century and rapidly became the largest consumer of dry cells. About two million batteries were sold and used in the United States alone during the year 1900.

The years following the First World War brought forth many devices requiring electrical power. All early forms of radio receivers were powered by batteries. It was not until the late 1920s that rectifiers and capacitors were developed and a radio could be powered by commercial ac power lines. Battery use was increasing and this spurred further interest in developing better and cheaper batteries.

The next major advance in batteries occurred in the early 1940s with the development of mercuric oxide alkaline cells by Samuel Rubin. This type of cell uses zinc as an anode material and mercuric oxide as the cathode material. The electrolyte is a paste of potassium hydroxide. This cell uses an alkaline (basic) electrolyte rather than an acidic electrolyte. The mercuric oxide cell was a vast improvement over the zinc-carbon types then available. The mercuric oxide cell packed more power into a smaller volume. It lasted longer and was less sensitive to temperature than the zinc-carbon type. This type of cell was used extensively by the military for portable communications, mine detectors, and flashlights. The mercuric oxide alkaline cell is the father of the modern *alkaline* dry cell.

During the 1970s a new type of battery was developed; the *lithium cell*. All commercial portable dry type primary cells to this point used zinc as an anode material. The lithium cell uses lithium metal as an anode. Cathode materials vary widely; thionyl chloride, sulphur dioxide, and manganese dioxide are all used. The electrolytes are usually alkalimetal salts dissolved in a non-water organic solvent. These cells are high-tech wonders producing about twice the power density of any other type of battery. They have the longest shelf life and are relatively less sensitive to temperature.

The history of commercially produced batteries is closely related to the development of manufacturing technology. For instance, Edison was working with crude forms of lithium cells as early as 1908. It was not, however, until the 1970s that manufacturing technology was able to deliver commercial models of the lithium cell. The commercially available primary cells represent only a small fraction of the possible electrochemical couples which could be used. Manufacturing technology and, above all, cost are primary factors determining which types of batteries are mass produced.

Manufacturing technology is responsible for many basic improvements in all types of batteries. These improvements are mechanical in nature; the basic chemical operation of the cell remains the same. The electrical parameters of the cell are basically determined by the active materials used in the cell. Improved manufacturing techniques have resulted in cells with higher power densities and longer lives.

All cells have one thing in common; they produce electricity by the oxidation of a metal. The metal used has a high affinity for oxygen and is the anode of the cell. In the case of primary cells, this anode metal is either zinc or lithium. The materials supplying the oxygen for reaction vary widely, with manganese dioxide being the most common. The oxygen supplying material is the cathode. Electrolytes, again, very widely but they all perform the same function. The electrolyte is a medium for electron transfer between the active materials.

In order to use this armada of available batteries wisely, it is necessary to examine the characteristics of each type of cell. Proper primary cell usage consists of choosing the right battery for the specific job. Each type of cell has its own particular advantages, which sometimes transcend cost. The cheapest cell to buy may not be the cheapest cell to use. The information given on each cell type will enable you to choose the proper battery to do the job in the most cost effective manner.

In the following material, energy density is expressed in two forms: by cell weight and by cell volume. Energy density in relation to cell weight is expressed as watt-hours per pound, and in relation to volume it is expressed as watt-hours per cubic inch. These figures give an idea of how much energy is available in relation to the cell's physical size and mass. The information on shelf life is based on the time it takes for the cell to discharge itself to an 85 percent state of charge. Shelf life figures assume that the cell is being stored, and not being discharged through an external load. Shelf life is greatly affected by temperature in all types of electrochemical primary cells.

ZINC-CARBON CELLS

The zinc-carbon cell is the oldest and most commonly used cell. It is a direct descendant of the Leclanché cell. The zinc-carbon cell is commonly called the *flashlight battery*, named for the device in which most of them are used. This battery represents a portable power source that is delivered at the absolute minimum in cost. The zinc-carbon cell is manufactured in all small sized dry cell cases. Some battery manufacturers do not identify them by the words *zinc-carbon*, but instead call them *general purpose* cells.

Chemical and Physical Construction

From a chemical viewpoint, the zinc-carbon cell is misnamed. The anode of the cell is indeed constructed of zinc (Zn) metal. The active material of the cathode, however, is composed of manganese dioxide (MnO_2). The manganese dioxide is the source of the oxygen necessary to react with the zinc and to thereby produce electricity. Carbon is used in the cathode only as a current collector; it does not actually enter into chemical change with the active materials of the cell.

The electrolyte in the zinc-carbon cell is ammonium chloride NH_4Cl) and zinc chloride($ZnCl_2$). The electrolyte is an acid and exists in the form of paste. The electrolyte from the common zinc-carbon cell is corrosive and can cause damage if it escapes from the cell enclosure. The corrosive nature of the electrolyte makes it impossible to use steel in the internal mechanical construction of the cell.

The zinc-carbon cell is contained within a cylindrical container of extruded zinc metal. The outside cell cover is merely polyethylene coated paper to prevent the cells from shorting when assembled into packs. The outside structural member, the zinc tube, is the case of the cell as well as the anode of the cell. The inside of the zinc tube is the surface area which reacts chemically as the anode. There is a separator, composed of flour, starch, and small amounts of mercuric chloride, which lines the inside of the zinc tube. The separator is saturated with electrolyte and provides a barrier to prevent direct circuits between the anode and the cathode.

The cathode material saturated with electrolyte resides inside the separator. The carbon rod is centered in the cathode material and performs the function of current collector. The carbon rod is mechanically attached to the top (positive pole) of the cell. The carbon rod rests on paper washers at the bottom of the cell. These paper washers prevent a direct short circuit between the case of the cell and the carbon rod. The case of the cell is the negative pole. The cathode material is held in place by an asphalt seal near the top of the cell. Above this seal there is an air space to minimize cell damage in the event of gassing. This air space is vented outside through a vent washer. Figure 5-2 illustrates the interior construction of the zinc-carbon dry cell.

The major problem with this type of cell construction is that the structural case of the cell is also the anode. The anode must enter into the chemical reaction in order to produce electricity. The zinc cell case is gradually consumed from the inside by chemical reaction. If the cell does not uniformly discharge, it is possible that the electrolyte will eat a hole in the zinc case and escape. The reason that

73

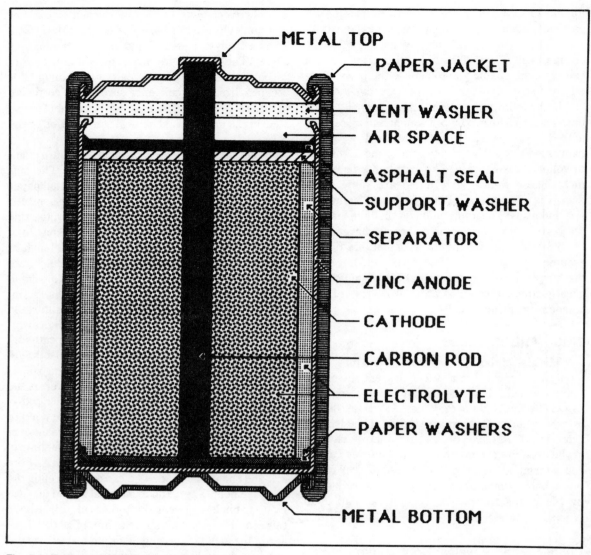

Fig. 5-2. Zinc-carbon cell internal construction.

the zinc container is also used as the anode is one of cost. It's simply the easiest and cheapest method of cell construction.

It is desirable to remove weak or dead zinc-carbon cells from the device they are powering. As these cells approach empty, the possibility of leakage greatly increases. Dead cells which remain in devices will eventually spill their electrolyte and ruin the device. This is especially true at temperatures above 85° F.

The *heavy duty* zinc carbon cell has the same type of construction. In the heavy duty cell, the electrolyte is pure zinc chloride and the cathode material is a higher grade of manganese dioxide. The use of pure zinc chloride as an electrolyte gives the cell a lower internal resistance. This lower internal resistance and more concentrated cathode material give the heavy duty cell the ability to deliver higher rates of discharge for longer periods of time than the standard zinc-carbon cell.

Energy Specifications

The zinc-carbon cell has a voltage of 1.5 volts. This voltage reflects the electrical nature of the zinc oxidation chemical change. All zinc-carbon cells exhibit the same voltage regardless of size, i.e. capacity. The heavy duty zinc carbon cells have the same 1.5 volt cell voltage. The voltage of the zinc-carbon cell drops gradually as the cell is discharged. The greater the discharge rate the faster the voltage drops off. Figure 5-3 illustrates the relationship between cell voltage and discharge time. The cell used here for an example is a standard zinc-carbon D-sized cell at room temperature. The information is derived from continuous discharge of the cell.

The zinc-carbon cells are available in a variety of sizes, from AAA to D. The AA cell has a capacity of about 0.9 ampere-hours, with the heavy duty version containing some 1.1 ampere-hours. The C-sized cell contains about 2 ampere-hours, with the heavy duty version containing about 2.6 ampere-hours. The D-sized cell contains about 4 ampere-hours in the standard model, with the heavy duty version containing about 6 ampere-hours. The small rectangular nine-volt battery is usually only manufactured in the heavy duty version and has a capacity of about 0.3 ampere-hours. These capacities are averages; the actual capacity of a cell depends on the specific manufacturer and the age and temperature of the cell.

The energy density of the standard zinc-carbon cell averages about 38 watt-hours per pound and 2.5 watt-hours per cubic inch. The heavy duty zinc-carbon cell has an energy density of 46 watt-hours per pound and 3 watt-hours per cubic inch. These figures are averages based on zinc-carbon cells produced by a number of major manufacturers.

Shelf Life

The shelf life of the zinc-carbon cell is between one and two years at room temperature. The shelf life of the cell is greatly effected by temperature.

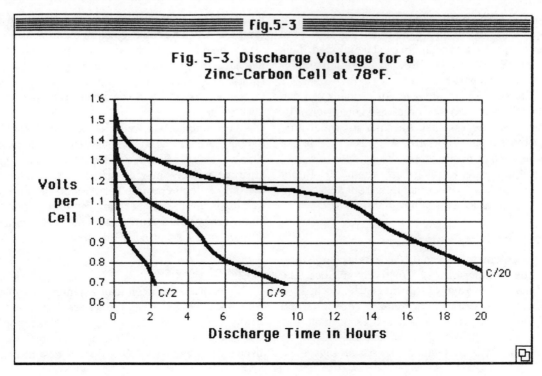

Fig. 5-3. Discharge Voltage for a Zinc-Carbon Cell at 78°F.

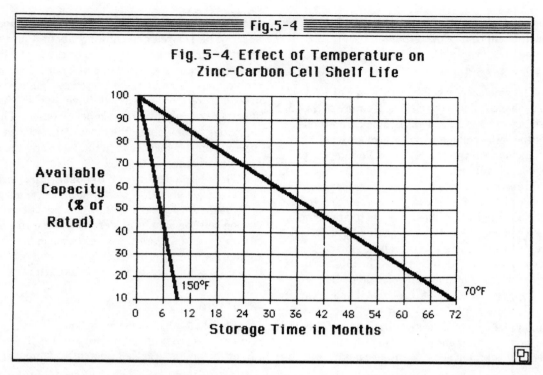

Figure 5-4 illustrates shelf life as a function of temperature for the zinc-carbon cell. The higher the temperature the less shelf life can be expected. This is due to electrolyte loss from evaporation and internal self-discharge.

The shelf life of any type of primary cell can be greatly extended by storage in a cool place. The zinc-carbon has less shelf life than any other type of primary cell. They are best used within six months of their manufacture. Prolonged storage of the cell at temperatures greater than 120° F. not only results in very short shelf life, but also can result in premature cell failure and leakage.

Effects of Temperature

In addition to increasing the self-discharge rate, extreme temperatures effectively reduce the capacity and voltage of the zinc-carbon cell. Low temperatures result in lowered cell capacities and voltages. If the zinc-carbon cell is used at low temperatures (below 32° F.) about one-half of the rated capacity is not available for usage. Figure 5-5 illustrates the relationship between cell voltage and temperature. The cell used as an example is a AA-sized heavy duty zinc-carbon dry cell being discharged at a C/6 rate (0.15 amperes). The response of standard zinc-carbon cells to low temperatures is even more severe than the heavy duty types.

The zinc-carbon cell is not suited for low temperature service. It has the greatest voltage and capacity loss in relation to temperature of any type of primary cell. The zinc-carbon cell is suited for use only at temperatures between 50° F. and 95° F. This is not to say that they cannot be operated outside this temperature range, but that is neither efficient nor cost effective to do so.

Types of Service

The zinc-carbon cell is only suited to applications with low discharge rates (less than C/20). In terms of a D-sized zinc-carbon cell, this amounts to a discharge current of 0.2 amperes or less. At discharge rates over C/20, it is more cost effective to use another type of primary cell. This is especially true if the battery is being continuously discharged. The zinc-carbon cell performs better in intermittent service than under continuous discharge conditions. This is due to its relatively high internal resistance.

The zinc-carbon cell is best suited to intermittent operation at low rates of discharge. Examples of this type of service are portable radios and flashlights when used intermittently. Actually, just about the only good factor of the zinc-carbon cell is its low cost. In most applications, it is more cost effective to use another type of cell.

Types of service that the zinc-carbon cells are not suited for are tape recorders/players, flashlights used for long periods, movie and VCR cameras, and radios with high-powered audio amplifiers. In general, all motorized devices do not perform well on zinc-carbon cells due to their high rates of current consumption. Zinc-carbon cells are not used in watches, calculators, and cameras due to the low energy density of the cell. They are simply too large for a given capacity to fit within the device. All these devices can be more cheaply and effectively powered by other types of primary cells.

Cost

Cost of the zinc-carbon cell is the lowest of any type of primary cell. Costs vary widely as these cells are commercially marketed by a very large number of outlets. In general, the cost of a standard zinc-carbon cell is about $ .10 per watt-hour. This is less than half the cost of the alkaline-manganese cell. This cost figure is remarkably cheap and reflects the extremely simple cell construction and low material cost.

In general, the larger cells cost less per watt-hour than the smaller sized cells. This is due to manufacturing and marketing costs. Consumers should beware of zinc-carbon cells for sale at reduced prices. They often are old stock. The short shelf life of the zinc-carbon cell makes old cells a poor bargain at any price.

ALKALINE-MANGANESE CELLS

The alkaline-manganese cell is commonly

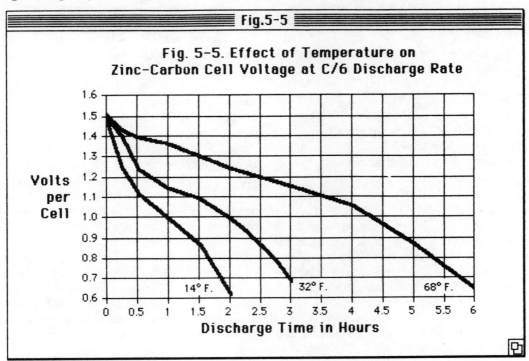

Fig. 5-5. Effect of Temperature on Zinc-Carbon Cell Voltage at C/6 Discharge Rate

called the *alkaline cell*. This type of cell represents an attempt to maximize the potential of the zinc-manganese dioxide chemical reaction. The alkaline-manganese cell contains higher quality active materials and an advanced physical construction to optimize the power output of the cell. The electrolyte of this cell is alkaline in nature and is responsible for the cell's name.

The alkaline-manganese cell is a very good example of the application of manufacturing technology to battery development. This cell uses the same basic chemical reaction used in the zinc-carbon cell. The reaction has been optimized by using sophisticated manufacturing techniques in cell construction. Every component of the alkaline-manganese cell is selected for maximum power output. There are no manufacturing shortcuts taken to save money. The price of these cells represents their greater energy densities and manufacturing costs.

Chemical and Physical Construction

The anode material of the alkaline-manganese cell is powdered zinc metal, amalgamated with small quantities of mercury. The anode material is carefully controlled for chemical purity and grain size. The anode material in the alkaline-manganese cell is not used as a structural material within the cell. Powdering the zinc metal gives more surface area available for chemical reaction. This area increase lowers the cell's internal resistance, resulting in higher energy density. The reduced internal resistance also results in far better cell performance under high discharge rates.

The cathode material is a mixture of manganese dioxide (MnO_2) and graphite. The manganese dioxide used in zinc-carbon cells is mined as a mineral from the earth. Purity of the natural product is not very high. The manganese dioxide used in manufacturing the alkaline-manganese cells is produced synthetically through electrolysis. Its purity and oxygen content are far superior to the naturally produced material. The higher oxygen content of the cathode material contributes to this cell's increased energy density and performance. Here again, extra cost is added to the cell to produce an electrically superior product.

The electrolyte of the alkaline-manganese cell is a solution of potassium hydroxide (KOH) in water. The electrolyte exists as a paste. This electrolyte is alkaline (basic) in contrast to the electrolyte of the zinc-carbon cell which is acidic. This difference allows steel to be used as a cell construction material. Potassium hydroxide is highly electrically conductive. The powdered, porous nature of the anode and the cathode materials allow them to be thoroughly saturated with the electrolyte solution. This even distribution of the electrolyte ensures that all the active materials within the cell are consumed during discharge.

The internal physical construction of the alkaline-manganese cell is radically different from the zinc-carbon cell. The alkaline-manganese cell is entirely encased in steel. This outer steel jacket has no function other than to support the cell. It is not used in the chemical reaction and is purely a structural component. The steel jacket provides a stronger and more secure container for the cell. The outside steel jacket is electrically insulated from the rest of the cell by a plastic sleeve.

Inside the plastic sleeve there is another steel container. This container is in intimate contact with the cathode material within it. This inside steel jacket acts as a current collector for the cathode materials in addition to containing them. The cathode materials are compressed within the inner steel jacket. The cathode material exists as a cylinder. Inside the cylinder of cathode materials there is a separator which is constructed of porous synthetic fibers. The separator provides electrical insulation between the anode and the cathode materials. The anode material resides inside the separator. The anode material, the separator, and the cathode material are all saturated with electrolyte solution.

Imbedded in the anode material is the anode current collector, often called the *nail* because of its shape. This metallic nail is in contact with the anode material and is welded to the bottom cap of the cell. The nail extends through a vented plastic

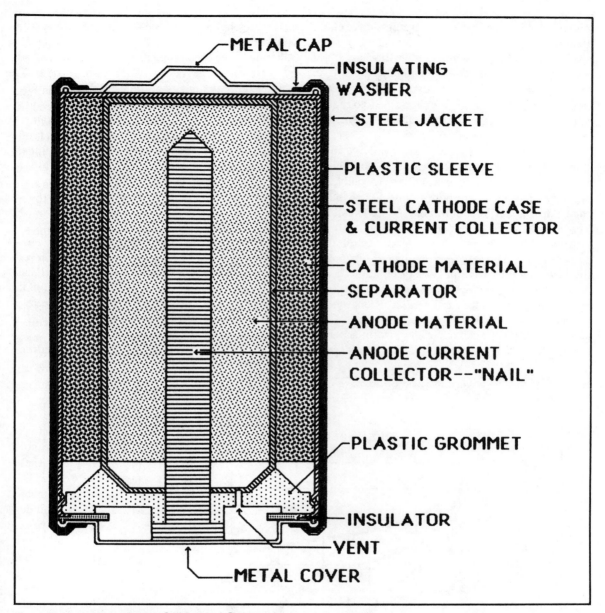

Fig. 5-6. Alkaline-manganese Cell Internal Construction.

insulator at the base of the cell. This vent allows for electrolyte expansion and protects the cell from rupturing. Figure 5-6 illustrates the internal construction of the alkaline-manganese cell.

Energy Specifications

The voltage of the alkaline-manganese cell is about 1.5 volts. This reflects the nature of the same chemical reaction used in the zinc-carbon cell. As with all electrochemical cells, the cell voltage of the alkaline-manganese cell is dependent on rate of discharge. Voltages while under discharge are much higher for the alkaline-manganese cell than the zinc-carbon cell, especially at high discharge rates (over C/20). Figure 5-7 shows cell voltage in relation to time for three discharge rates. The temperature is assumed to be 68° F.

Alkaline-manganese cells are available in all standard dry cell sizes. They are also available in a wide variety of button cases. The AA cell has a capacity of 1.5 ampere-hours. The C cell has a capacity of about 5 ampere-hours. The D cell has a capacity of about 10 ampere-hours. The small rectangular 9 volt battery has a capacity of 0.45 ampere-hours. These figures are averages based on the cells made by several battery manufacturers. Actual usable capacity is dependent on battery age, operating temperature, and rate of discharge.

The relatively low internal resistance of the alkaline-manganese cell makes it perform superbly at high rates of discharge. At discharge rates over C/10 the alkaline-manganese cell outperforms its zinc-carbon relative by factors greater than two. Alkaline-manganese cell construction assures that even rapidly cycled cells will not rupture and leak.

The energy density of the alkaline-manganese cell is 45.5 watt-hours per pound and 3.9 watt-hours per cubic inch. The energy density of the alkaline-manganese cell is greater than the zinc-carbon type. Of all the commonly marketed zinc-manganese dioxide cells, the alkaline-manganese cell packs the most power within the smallest size.

Shelf Life

The shelf life of the alkaline-manganese cell is between two and three years if the cell is stored at room temperature. After this period, the cell has lost 15 percent of its rated capacity. Figure 5-8 illustrates the relationship between capacity retention and time. This graph shows three curves for three different storage temperatures.

This information tells us that alkaline-

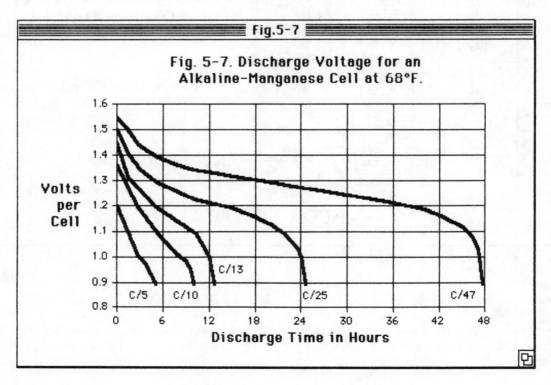

Fig. 5-7. Discharge Voltage for an Alkaline-Manganese Cell at 68°F.

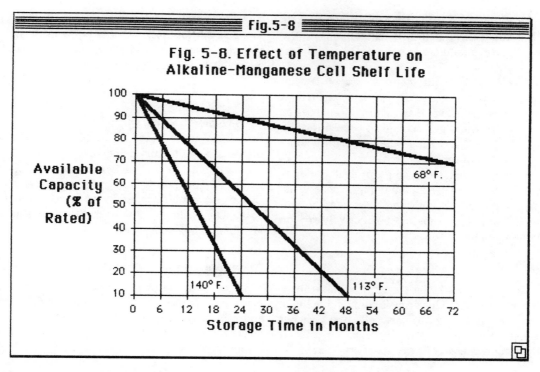

Fig. 5-8. Effect of Temperature on Alkaline-Manganese Cell Shelf Life

manganese cells are best stored at cool temperatures. At temperatures over 120° F. the shelf life of the cell is reduced to about eight months. The alkaline-manganese cell is very tightly sealed and is less prone to electrolyte loss from evaporation. The alkaline-manganese cell has a much longer shelf life than the zinc-carbon cell.

Effects of Temperature

Temperature affects the voltage of the alkaline-manganese cell. Figure 5-9 illustrates the relationship between cell voltage and temperature for the alkaline-manganese cell under discharge at the C/13 rate. The alkaline-manganese cell is less affected by temperature than its zinc-carbon relative. The good high and low temperature performance of this cell is due to its very low internal resistance (typically less than one Ω in alkaline cells).

The alkaline-manganese cell is useful at temperatures between 0° F. and 138° F. They can be used at temperatures outside this range. However, they will suffer voltage and capacity loss. The extremely rugged mechanical cell construction allows the alkaline-manganese cell to operate over a wide temperature range without cell failure and leakage.

Types of Service

Alkaline-manganese cells are suited for discharge rates as fast as C/2. This type of cell is generally the most cost-effective type to use in all disposable battery applications. The alkaline-manganese cell's long shelf life and mechanical security make this cell highly desirable for expensive and seldom used devices. The high energy density of the alkaline-manganese cell makes it very popular in many portable devices.

Flashlights draw between 0.2 and 0.6 amperes from their cells when operating. This is a fast discharge rate, even for a cell as large as the D cell. Alkaline-manganese cells are better suited to these discharge rates than the zinc-carbon type. This is especially true if the flashlight is under continuous operation. For example, if a flashlight using two D cells is continuously operated, its zinc-carbon cells will last about three hours. If the cells inside the flashlight are alkaline-manganese they will last about 18 hours.

Portable audio equipment should be run on alkaline-manganese cells. This is especially true if a motor is involved, such as in a tape recorder or player. If the audio gear is high powered (over 5 watts), then alkaline cells will outperform zinc-carbon cells by a factor of four.

Any portable device using an electric motor should be powered by alkaline-manganese cells. Devices such as toys and motorized cameras are examples which are most cost-effectively run on alkaline-manganese cells.

Cost

The cost of the alkaline-manganese cell is about $ .13 per watt-hour. This price is comparable to the zinc-carbon cell. As with most cells, the cost per watt-hour is greater in the smaller sized cells. If alkaline-manganese cells are discharged at rapid rates, they offer much less cost per watt-hour than the zinc-carbon type. Actually, considering its manufacturing cost and complexity, the alkaline-manganese cell is remarkably inexpensive.

MERCURY CELLS

The mercury cell, which has been around since the 1940s, is now finding its place in consumer markets. The mercury cell in its *button* or *disk* case is used universally in watches, calculators, hearing aids, and cameras. It is not so commonly used in the larger standard cases due to its relatively high cost. This cell is actually a forerunner of the alkaline-manganese type.

Chemical and Physical Construction

The anode material is composed of highly pure zinc metal. The anode material is in powdered form and tightly compressed. The cathode material is mercuric oxide (HgO), which is also powdered and compressed. Mercuric oxide is a highly concen-

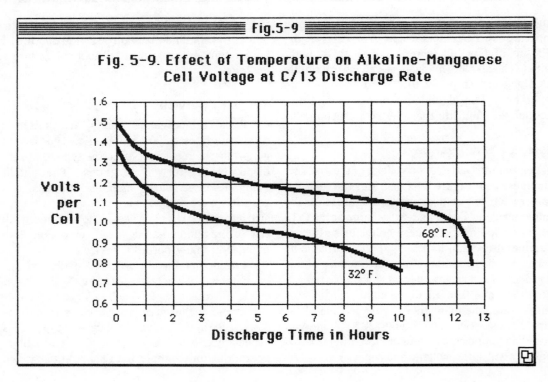

Fig. 5-9. Effect of Temperature on Alkaline-Manganese Cell Voltage at C/13 Discharge Rate

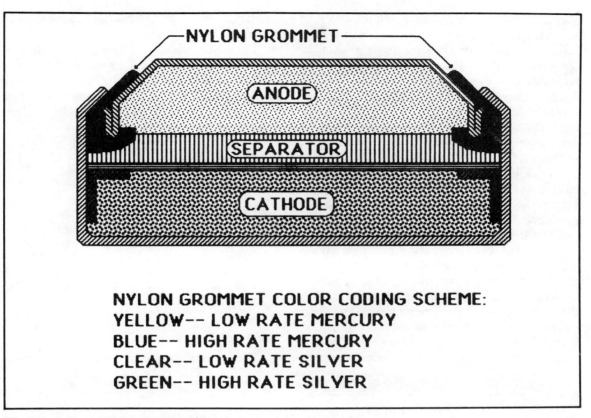

Fig. 5-10. Button cell internal construction.

trated source of oxygen for reaction with the anode. It is, however, many times more expensive than manganese dioxide as a cathode material.

The electrolyte used in the mercury cell is a paste of potassium hydroxide (KOH) in water. This electrolyte is optimized for high rates of discharge. In mercury cells designed for lower discharge rates, the electrolyte used is a paste of sodium hydroxide (NaOH) and zinc oxide (ZnO) in water.

The internal mechanical construction of the larger mercury cells is very similar to that of the alkaline-manganese cells. The mercury cell is most commonly used in the *button case*. This case, while different in form, has all the same standard design elements which are merely rearranged in a smaller package. Figure 5-10 illustrates the interior construction of the button cell. The button cell is a very rugged mechanical assembly and leakage is extremely uncommon with this type of case.

One interesting feature of the button case is the color coded nylon grommet which seals and insulates the top of the cell. In mercury cells designed for high discharge rates, the nylon grommet is colored blue. In low discharge rate mercury cells this grommet is colored yellow.

The case of a mercury cell is a precision made item. The additional cost of the mercuric oxide warrants the extra attention paid to the mechanical details of cell construction. The steel case is plated with nickel and in some instances gold and silver to facilitate electrical contact. The mechanical seal of the case is very tight. Leakage is virtually unknown.

Energy Specifications

Cell voltage of the open-circuit mercury cell is

about 1.4 volts. The discharge voltage curve remains fairly constant over the entire capacity of the cell. Due to its relatively low internal resistance, the mercury cell's voltage is not as affected by the rate of discharge as some other types of cells. Figure 5-11 shows the cell voltage of a mercury cell in relation to time for four discharge rates. The voltage stability of the mercury cell makes it ideal to use in requirements which demand constant voltages.

The mercury cell is at its best when discharged at rates less than C/50. Rates of discharge in this neighborhood reflect usage in a device such as a calculator. Watches have rates in the neighborhood of C/600 and are prime users of mercury cells. Shelf life and energy density are prime factors in devices which use energy at very slow rates.

The mercury cell is available in a large number of different capacities. It is manufactured in every size from micro-buttons to the standard D-sized dry cell package. A button cell will have a capacity between 0.06 and 3 ampere-hours. A AA-sized package will have a capacity of about 1.8 ampere-hours. A D-sized mercury cell has a capacity of 16 ampere-hours.

The energy density of the mercury cell is 58 watt-hours per pound and 6.4 watt-hours per cubic inch. The mercury cell packs about three times the power of a zinc-carbon cell in the same space. This high energy density reflects the tremendous oxygen potential of the mercuric oxide.

Shelf Life

The shelf life of a mercury cell stored at room temperature is in excess of two years. As with all electrochemical cells, high temperatures accelerate cell decay. A mercury cell stored at 120° F. has a shelf life of about six months. Figure 5-12 illustrates the relationship between mercury cell shelf life and temperature. Remember, shelf life time is rated when the cell has lost 15 percent of its rated capacity.

Effects of Temperature

The mercury cell operates best at room

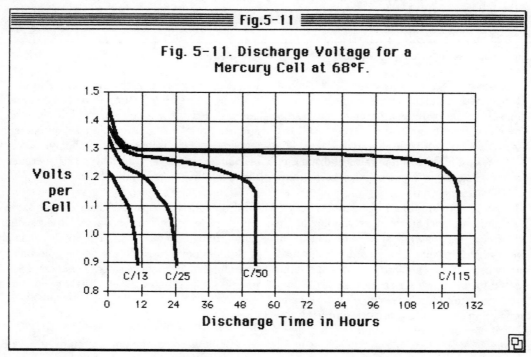

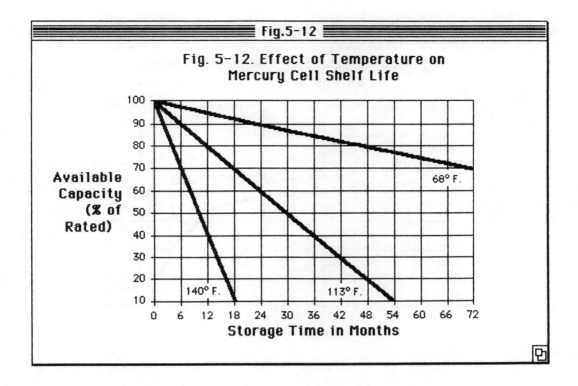

Fig. 5-12. Effect of Temperature on Mercury Cell Shelf Life

temperatures. It is capable of operation over a range of temperatures between 20°F. and 120°F. At temperatures outside this range both voltage and capacity are reduced. Figure 5-13 illustrates the effect of temperature on mercury cell voltage for a discharge rate of C/50.

Types of Service

The mercury cell is primarily used in applications that require small batteries with high volumetric energy density. The high cost of this type of cell restrict its usage to the smaller sizes. Usage in devices such as flashlights and other high drain devices is not cost effective.

Some examples of mercury cell usage are watches, hearing aids, exposure meters, calculators, and wireless microphones. The mercury cell is also used as a voltage standard for various measuring devices. This application is possible because of the very constant voltage output on the mercury cell when discharged at very slow rates.

Cost

The mercury cell is relatively expensive, costing between $1.20 and $3.40 per watt-hour. The actual cost of any given mercury cell is highly dependent on the size of the cell. Mercury cells are manufactured in such a wide variety of cases that it is impossible to generalize on cost figures. As with other cells, the larger the cell size the less the cost per watt-hour of the energy contained within it. When a consumer purchases a cell, he is not only buying the chemicals within the cell, but also the manufacturing and marketing effort it took to make the cell and to get it to him. In the smaller cells, the manufacturer and distributor spend about the same to market a small button cell as a large one. This makes the smaller sized cells more expensive in terms of dollars per watt-hour.

The high cost of the mercury cell reflects the additional cost of the mercuric oxide and the sophisticated manufacturing processes used. Mercury bearing compounds are not very plentiful, and

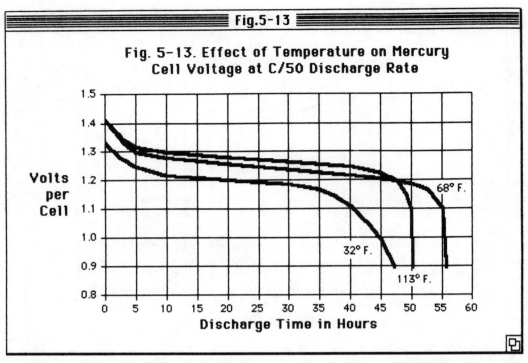

Fig. 5-13. Effect of Temperature on Mercury Cell Voltage at C/50 Discharge Rate

it is highly unlikely that mercury cell prices will be reduced in the future.

SILVER OXIDE CELLS

The silver oxide cell is a very close relative of the mercury cell. Chemically they are very similar and serve in many of the same applications. The major differences between these two types of cells are shelf life, volumetric energy density, and cost.

Chemical and Physical Construction

The anode material of the silver oxide cell is compressed and powdered zinc metal. The cathode material is powdered and compressed silver oxides. The cathode may be composed of one of two chemical forms of silver oxide—silver oxide (Ag_2O) and silver peroxide (Ag_2O_2). The performance characteristics of these two forms of silver oxide vary slightly. The extra oxygen content of silver peroxide gives cells made from it greater energy densities.

The electrolyte used in the silver oxide cell depends on the type of service the cell is designed for. For high rates of discharge, the electrolyte is a paste of potassium hydroxide (KOH) in water. In cells designed to be used at lower rates of discharge, the electrolyte is a paste composed of sodium hydroxide (NaOH) and zinc oxide (ZnO) in water.

The silver oxide cell is mechanically constructed in the same fashion as the mercury cell. The button case is the most common form of the silver oxide cell. Figure 5-10 illustrates the interior construction of a button cell. Due to the relatively high cost of silver oxide cells, they are not commonly manufactured in the larger sized cases.

As with all types of button cells, the materials composing the cell and the discharge rate are color coded in the cell's nylon grommet. The silver oxide button cell designed to be discharged at high rates has a green nylon grommet. The low rate silver oxide button cell has a clear grommet. This feature makes it easy to tell at a glance what type of cell is being used.

Energy Specifications

The silver oxide cell has a nominal voltage of 1.55 volts. At the very low rates of discharge for which the silver oxide cell is designed, this output voltage is extremely stable. Figure 5-14 illustrates the relationship between cell voltage and time for three discharge rates. These discharge rates are all relatively slow, in keeping with the types of service in which silver oxide cells are generally applied.

Silver oxide cells are available in capacities between 0.025 to 0.200 ampere-hours. These button cells are optimized for discharge rates around C/1000. In terms of current, these cells are discharged between 0.00005 and 0.003 amperes. The high cost of the silver oxide cells makes them unsuitable for competition with other types in the larger sized packages.

The silver oxide cell using Ag_2O as a cathode material has an energy density of 62 watt-hours per pound and 9.1 watt-hours per cubic inch. The silver oxide cell using Ag_2O_2 as a cathode material has an energy density of 73 watt-hours per pound and 10.8 watt-hours per cubic inch. The silver oxide cell, regardless of which cathode material is used, has the greatest volumetric energy density of any type of cell. The silver oxide cell packs the most power into a given space. This is true even if the new lithium cells are considered.

Shelf Life

The silver oxide cell really shines when it comes to shelf life. At room temperature, this type of cell has a shelf life of well over three years. Like most cells, its shelf life is affected by the temperature at which the cell is stored. Figure 5-15 illustrates the relationship between shelf life and temperature for the silver oxide cell. As with all types of batteries, the silver oxide cell is best stored at cool temperatures.

Effects of Temperature

The output voltage of the silver oxide cell is not as dependent on temperature as most other

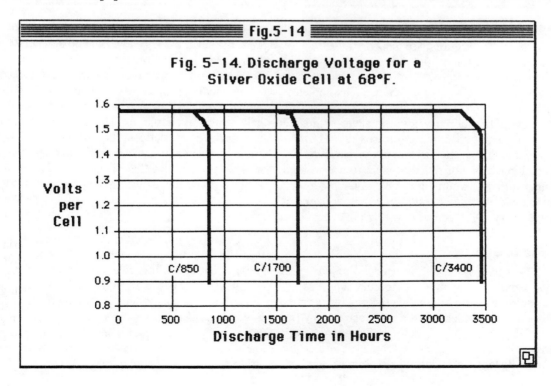

Fig. 5-14. Discharge Voltage for a Silver Oxide Cell at 68°F.

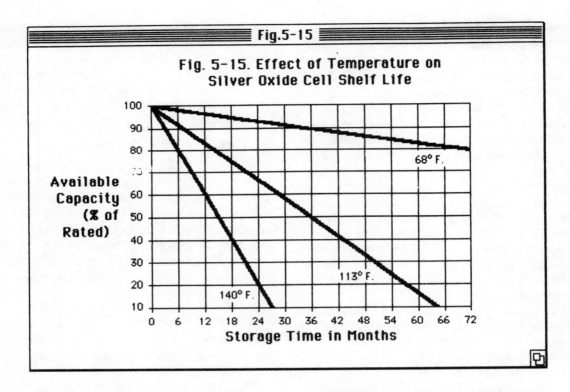

Fig. 5-15. Effect of Temperature on Silver Oxide Cell Shelf Life

types of cells. Figure 5-16 shows the effect of temperature on the output voltage of a silver oxide cell. The discharge rate is C/3400, which is very slow and reflects usage in a device such as an electronic watch.

As with the mercury cell, the silver oxide cell operates best at room temperatures. It is, however, capable of operation at temperatures over 150° F., making it one of the best cells to use in high temperature environments. The silver oxide cell has the widest range of operating temperatures of any primary cell discussed to this point.

Types of Service

The silver oxide cell is used in many of the same applications as the mercury cell. Many devices which were originally powered by mercury cells can be powered more cost effectively by the silver oxide cell. The greater volumetric energy density of the silver oxide cell makes it less costly to use if its long life is considered. In most applications, the silver oxide cell will last 50 percent longer than a mercury cell of the same size.

Some applications require silver oxide cells due to their small size in relation to their capacity. Ultra-miniaturized devices such as hearing aids have very little space available for the battery. In many miniaturized applications, the battery is the largest single component in the device. Silver cells are used here simply because there is nothing else that can pack as much power into a very small space.

Cost

The silver oxide cell is one of the most costly types in common usage. This stands to reason; silver is not cheap regardless of its form. Cost for the silver oxide cell is around $3.00 per watt-hour. Even though the silver cell is initially more expensive to purchase, it may be cheaper to use in low drain applications. The higher energy density and longer life make the silver cells more cost effective in electronic watches and calculators.

ZINC-AIR CELLS

The zinc-air cell is currently being marketed in Europe and is unavailable in the United States. This is a novel and still experimental type of electrochemical cell. Since all primary batteries discussed to this point make electricity by the oxidation of zinc, why not use the ambient air as an oxygen source? Experimental primary zinc-air cells show great promise in the consumer market.

Chemical and Physical Construction

The anode of the zinc-air cell is composed of zinc metal. The cathode material is the oxygen in the ambient air surrounding the cell. The electrolyte is a paste of potassium hydroxide (KOH) in water.

The usage of ambient air has caused problems with cell contamination and premature cell failure. This problem is inherent in the cell's design. The ambient air is not always clean. It is unlikely that this type of cell will ever be used in dusty or smoky environments.

The zinc-air cell has its air intake opened when it is placed in service. This allows the oxygen in the air to contact the separator within the cell. Here the oxygen reacts with the zinc through the electrolyte. Once a zinc-air cell is activated it cannot be shut off. The cell remains operational until it is exhausted.

Energy Specifications

The zinc-air cell has a voltage of 1.4 volts. The capacities of these cells vary from about 0.2 to 0.4 ampere-hours for a hearing aid button cell. Due to the experimental nature of this type of cell there is no hard information available on overall cell performance.

The energy densities of the zinc-air cell are estimated to be 141 watt-hours per pound and 7.7 watt-hours per cubic inch. The energy densities of this cell promise to be so high because of the absence of a cathode material within the cell. The cathode material is air, supplied from the outside. This makes the cell very compact and hopefully inexpensive.

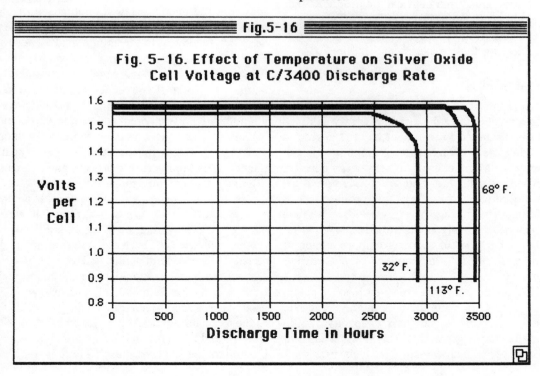

Fig. 5-16. Effect of Temperature on Silver Oxide Cell Voltage at C/3400 Discharge Rate

Shelf Life

There is little hard data on zinc-air cell shelf life. They simply haven't been around long enough to know how long they can be stored. If their cell remains sealed, it is projected that the zinc-air cell will have an indefinite shelf life. Temperature does not effect the shelf life of an inactive zinc-air cell. This feature is one reason that battery manufacturers are doing research on this cell.

Types of Service

The current prototypes of the zinc-air cell have been designed for use in hearing aids. If the zinc-air technology can be perfected, then there should be a large market for the cells based on their long shelf life alone. If devices are designed with vents so the cell can breathe, then the zinc-air cell could find a large variety of applications .

Cost

The cost of the zinc-air cell is as yet undetermined. It is hoped that cells can be produced that deliver electricity for as little as $ .08 per watt-hour. Current cell prototypes cost many times this figure, reflecting their experimental nature.

LITHIUM CELLS

The lithium cell is a relatively new type of cell that is revolutionizing the primary battery field. All the cells discussed to this point have used zinc as an anode material. The usage of lithium as the anode material offers cells of higher energy density, much longer shelf life, and greater temperature operating range. Lithium was first proposed as an anode material during the early 1900s. The first practical commercial models were developed in Japan in the early 1970s. Commercial production of the lithium cell is a triumph of manufacturing technology.

Lithium has the greatest electrochemical potential of any known metal; it is highly reactive. Lithium has a theoretical energy capacity of 3.86 ampere-hours per gram; this is higher than any other metal. This factor results in cells of higher voltages and very low internal resistances (typically less than 0.6Ω). Lithium is also a very light metal, with a density about 1/3 that of other anode materials. This makes the cells weigh very little. Lithium is the ideal anode material for primary electrochemical cells.

Lithium cell technology is being pursued by many different battery manufacturers worldwide. Each manufacturer is developing a slightly different form of the lithium cell. Over 30 different chemical combinations of lithium cells have been produced. It looks like there may be as many as 100 different chemical combinations that can result in workable lithium cells. The lithium cell is currently manufactured for commercial marketing in five different chemical combinations. Each is slightly different from the next. However, all lithium cells are radically better performers than any other type of primary cells.

Chemical and Physical Construction

All lithium cells have one chemical element in common: the anode is always composed of high purity lithium metal. This lithium metal anode may be either solid or in the form of a grid . There are significant technical problems involved in the handling of lithium metal. It is highly reactive and will oxidize rapidly in the presence of either water or oxygen. The air surrounding us contains large amounts of both oxygen and water vapor. The lithium cell must be manufactured in a water free and oxygen free atmosphere. This greatly complicates the manufacturing process and makes the lithium cells currently more expensive.

It is in the choice of cathode material and electrolyte that the many types of lithium cells differ. All types use the oxidation of the lithium anode to produce electricity. There are, however, many different chemical compounds that can be used as cathode materials to supply this material for oxidation. In fact, the chemical element oxygen is not used in some cells which used cholorine or fluorine as the oxidizing agent. Add to this a variety of electrolytes and there is a very large number of possible lithium cells. The discussion here is restricted to now commercially available lithium systems. These systems were chosen because they will eventually replace zinc systems in most applications.

The differing electrolytes used in lithium cells all share one thing in common—they contain no water. In some cases the electrolyte is a solid. In other lithium cells, liquid organic electrolytes are used. In other cells the liquid electrolyte is inorganic. In all cases, the electrolyte contains no water, which is another radical departure from all primary cells discussed to this point.

Thionyl chloride ($SOCl_2$) is a cathode material that is in current use in lithium cells. The electrolyte used is also thionyl chloride. The usage of thionyl chloride as a cathode material results in a cell with very high output voltage. It probably has the highest output voltage of any commercially produced lithium cell.

Another cathode material used is poly-carbonmonofluoride ($[CF]_n$). The electrolyte used with this cathode material is an alkalimetal salt dissolved in an organic solvent. This chemical combination produces cells which are very stable chemically and generally non-toxic if the cell is breached. These cells are not internally presurized; hermetic sealing is not necessary. These cells are rugged with very wide temperature operating ranges. The poly-carbonmonofluoride cathode has probably the best chance of becoming a popular commercial success.

Sulphur dioxide (SO_2) is another cathode material that has been in use for about 12 years. Early prototype lithium cells were developed around the sulphur dioxide cathode. These early cells used sulphur dioxide gas dissolved under pressure in acetonitrile as an electrolyte. These cells were highly pressurized and leakage, while not common, did occur. Since the chemicals within this cell are dangerous, this type of cell is falling into disuse.

Manganese dioxide (MnO_2) is also used as a cathode material. These non-pressurized cells are popular in small button cases. Most of them see service in memory backup applications in computers. The manganese dioxide cathode does not produce the output voltage and energy density of other materials. This factor is limiting the production of lithium-manganese dioxide cells to relatively small sizes.

Iodine is another cathode material. The lithium-iodine cell has a relatively high internal resistance when compared to other types of lithium cells. This fact restricts its usage to very low drain applications. It is very long lived and is used extensively in cardiac pacemakers.

The mechanical constructions of these different types of lithium cells vary widely. They are in all cases very well made cells. The cases are constructed of nickel-plated steel or stainless steel, with all joints welded. This results in a tightly sealed case and is a contributing factor in this cell's long shelf life.

Energy Specifications

The cell voltages of the different types of lithium cells vary between 2.6 and 3.6 volts. This voltage variation is due to the different choices of cathode materials and electrolytes. Figure 5-17 illustrates the discharge curves for the lithium-thionyl chloride cell. Figure 5-18 shows the discharge curves for the lithium-poly-carbonmonofluoride cells. These curves illustrate the incredible performance of the lithium based primary cells. The discharge curves are very flat. The lithium cell maintains a very constant voltage at much higher discharge rates than any other type of cell.

The fact that most lithium systems have output voltages around 3 volts is a highly desirable feature. This is double that of any type of zinc primary system. This means that fewer cells need to be series interconnected to produce any given voltage. This results in smaller batteries with the same electrical capacity in watt-hours. This higher output voltage, coupled with lithium's light weight, combine to form a cell of the greatest energy density of any type of commercial battery.

The capacities of the various lithium cells cover a wide range between 0.025 to 8000 ampere-hours. Lithium cells are commercially available in all common sized dry cell packages. The AA-sized lithium-thionyl chlorine cell has a capacity of about

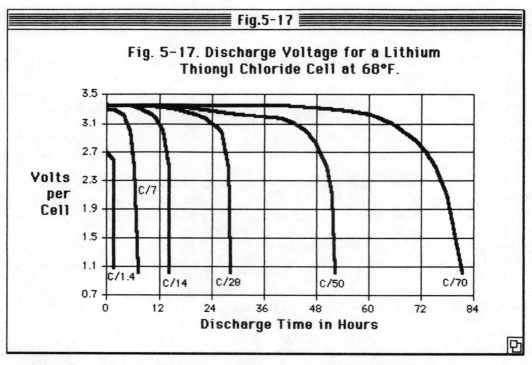

Fig. 5-17. Discharge Voltage for a Lithium Thionyl Chloride Cell at 68°F.

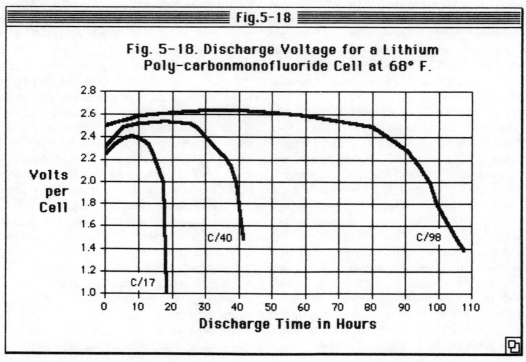

Fig. 5-18. Discharge Voltage for a Lithium Poly-carbonmonofluoride Cell at 68° F.

2 ampere-hours. The C-sized lithium-polycarbonmonofluoride cell has a capacity of 5 ampere-hours. The D-sized lithium-thionyl chlorine cell has a capacity of about 14 ampere-hours. These ampere-hour figures are higher than any other type of cell. These figures are based on current. If the additional voltage of the lithium cell is also considered, then the watt-hour rating of the cell is higher still. The lithium cell is an amazing breakthrough in the world of batteries.

The theoretical energy density of the lithium cell is very high, around 1000 watt-hours per pound. Commercially produced models now average around 160 watt-hours per pound. This is over three times that of the alkaline-manganese cells. Currently made lithium cells have a volumetric energy density of around 12 watt-hours per cubic inch. Lithium technology is just beginning. It is possible that cell performance will radically improve in the near future. Even at the energy densities being produced now, the lithium cell outperforms all other types of primary cells.

Shelf Life

The shelf life of currently manufactured lithium cells is better than 10 years. This figure is not greatly affected by temperature. The lithium cell has the longest shelf life of any type of battery over three times that of the next longest lived type.

One might think that a cell composed of such an active element as lithium would be unstable and have a very short shelf life. This is not the case. When the cell is first assembled there is a reaction between the anode and cathode which oxidizes the lithium metal. This oxide layer insulates the lithium metal anode from further self-discharge until the cell is used.

Effects of Temperature

The lithium cell's performance over a very wide range of temperature is nothing short of tremendous. These cells can effectively be used between the temperatures of −40°F. to 150°F. The voltage and capacity of the cell is minimally affected

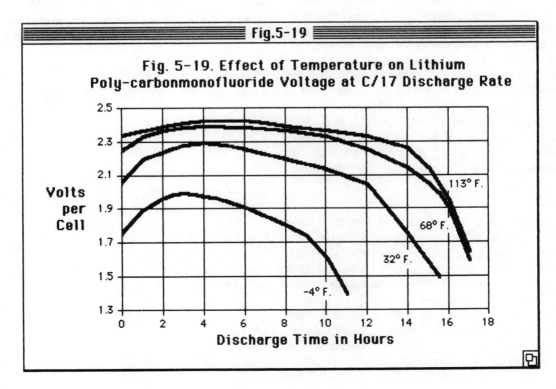

Fig. 5-19. Effect of Temperature on Lithium Poly-carbonmonofluoride Voltage at C/17 Discharge Rate

by commonly encountered temperatures. Figure 5-19 shows the effect of temperature on a lithium cell's voltage. The rate of discharge is C/17.

The capacity of the lithium cell is not greatly affected by temperature. These cells retain 90 percent of their rated capacity when used at temperatures as low as $-15°$ F. and as high as $125°$ F. These cells are very suitable for use in extreme environments.

Types of Service

Lithium cells are suitable for use in any and all types of portable electrical equipment. They are manufactured in sizes that will fit any particular requirement. They are not commonly used today due to their high expense. Their additional voltage makes it difficult to use them in battery holders designed to contain zinc cells. Lithium cells can be used in a flashlight if the bulb is replaced with one designed to run on the higher voltage.

The most common consumers of currently produced lithium cells are computers and the military. Lithium cells are commonly soldered to computer circuit boards as long-lasting backup power supplies for the computer's memory. Military uses are radio transceivers, torpedo batteries, guided missiles, telemetry systems, and laser sighting devices. In the future, we may expect to see lithium cells in everything from flashlights to super-sophisticated electronics. But first the lithium cell's price must come down. The higher voltage of these cells makes it necessary for device manufacturers to design their usage into the device's battery holder.

Cost

The current commercially marketed versions of the lithium cell cost an average of $1.97 per watt-hour. This price is currently competitive with mercury and silver oxide cells, but still much higher than the cost of alkaline-manganese cells.

The high cost of the lithium cell is a function of its manufacturing complexity. The actual raw materials that make up the lithium cells are not very expensive. Free lithium is not encountered naturally because it is so chemically reactive. Lithium compounds do, however, commonly occur in most igneous rocks. The dry lake beds of California and Nevada contain very large quantities of lithium salts. The basic raw materials used in lithium cells are cheap and available. The problem is in the handling of the pure lithium metal. All manufacturing processes must be carried on in a very dry inert atmosphere. As such, startup cost for lithium cell manufacture is very high. If mass production technology can be successfully applied to the lithium cell, then its cost will drop radically.

It is estimated that if the lithium cell's cost falls below $ .50 per watt-hour, it will capture a highly significant share of the primary battery market. While this figure is still about 5 times that of alkaline-manganese cells, the far greater performance of the lithium cell outweighs its additional cost. Due to the significantly better performance inherent in lithium cells, it is simply a matter of time until they replace all other forms of primary cells. These cells also offer advantages to the manufacturer. For the first time, a single chemical cell technology is superior in all phases of battery performance. This allows manufacturing procedures and machines to be optimized for this single lithium based cell technology.

PRIMARY CELL MANUFACTURERS

Duracell Inc.
Berkshire Industrial Park
Bethel, CT 06801
203-796-4155

Union Carbide Corp.
Danbury, CT 06817

Panasonic Industrial Company
Post Office Box 1511
Secaucus, NJ 07094
201-348-5266

Altus Corporation
1610 Crane Court
San Jose, CA 95112
408-295-1300

Eagle-Picher Industries, Inc.
Post Office Box 130
Seneca, MO 64865
417-776-2256

SAFT America, Inc.
107 Beaver Court
Cockeysville, MD 21030
301-666-3200

Varta Batteries Inc.
150 Clear Brook Road
Elmsford, NY 10523
914-592-2500

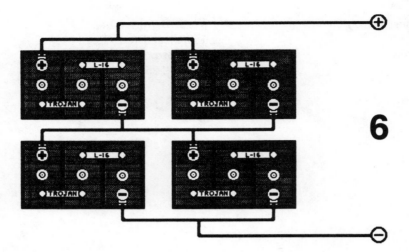

Methods and Machines to Charge Batteries

There are many methods of charging batteries. The method to be used is determined by the type of cells being charged and by their capacity. Large chargers are required for large storage batteries and small chargers for small flashlight type cells.

In addition to size there is the question of what sort of control is placed on the charging process. Voltage limitation is one type of control; current limitation is another. Some charging machines use both types of control, each at the appropriate stage of the charging process.

The methods and machines described in this chapter are primarily aimed at charging lead-acid cells of large capacity, and small nickel-cadmium cells. These different types of cells have very different charge characteristics and should not be confused. A thorough understanding of Chapters 2 and 3 will provide the information necessary to choose the proper charging method and machine for the job.

LEAD-ACID BATTERY CHARGERS

The lead-acid battery can be charged using a wide range or rates of charge. You can rapid-charge them in just a few hours or you can trickle-charge them for several days. In deep cycle service, the charge source should be able to provide at least a C/20 rate into the battery. Batteries in float service can be charged at rates as low as C/50.

The lead-acid cell is usually charged using voltage control to limit the rate of charge when the cells are full. The voltage limitation should be adjustable to allow periodic equalilzation of the individual cells within the battery. Amperage control is also very desirable in the beginning of the charge cycle. Amperage control prevents the cells from being charged too rapidly when they are empty.

NICKEL-CADMIUM BATTERY CHARGERS

The nickel-cadmium cell is charged using current limitation throughout its entire charge process. The ni-cad does not charge efficiently or safely using voltage limitation as the only control of the charging process.

The methods of ni-cad charging usually involve filling them at a fixed current rate for a certain period of time. If this method is used, care must be taken not to charge the batteries for too long a period of time. Don't put them in the charger and forget them. If the charger is supposed to fill them in 12 hours, then any charging over 24 hours in duration will shorten the cell's life.

120 VOLT AC POWERED CHARGERS

Commercial power is one of the most common energy sources used to charge batteries. Ac powered chargers are used to charge car batteries, large storage batteries, lead-acid gel cells, and ni-cads of all sizes and types.

The ac powered charger consists of a transformer to reduce the voltage of the commercial power to a lower voltage that is acceptable to the batteries. After the energy is transformed down it is rectified from alternating current (ac) to direct (dc). This dc current is then fed to the battery, usually through a control circuit.

Current Capacity

If an ac powered charger is used to charge a battery, then it must be capable of delivering enough energy to effectively do the job. Lead-acid cells are best charged at the C/20 rate. Ni-cads are best filled at rates between C/10 and C/20.

For example, if a 100 ampere-hour lead-acid car battery is to be charged, then we should use a charge source with at least a 5 ampere output (100 ampere-hours / 20 hours = 5 amperes). An output current of 10 amperes is acceptable as long as the battery is not charged for long periods of time. If a ni-cad with a capacity of 0.5 ampere-hours is to be charged, then the charge source must be capable of delivering a minimum of 0.025 amperes. Any charger used must have enough current output to properly charge the battery.

The current output of a charger is not a big concern in charging small ni-cads as the current requirements are so low. If the cells, however, have a large capacity, over 200 ampere-hours, then the tendency is to use smaller chargers due to the expense of larger output equipment. Batteries of large capacity need large charge currents. Don't buy a small charger and expect it to fill a large capacity battery.

Ac powered chargers for 12-volt lead-acid batteries are available everywhere. They are common because of the automobile starting battery. Trickle chargers that can supply about 2 amperes cost around $10. Chargers that have current capacities of 10 amperes cost around $50. Chargers that can deliver 30 amperes or more for hours at a time are expensive, $100 to $500.

The rated amperage output is measured with the charger delivering energy to a totally discharged battery. A charger that is rated at 10 amperes will deliver 10 amperes into a totally flat battery. When the battery has reached a 25 percent state of charge, the output from the charger will drop to about 7 amperes. At a 50 percent state of charge, the charger will deliver around 5 amperes. As the battery approaches 80 percent state of charge, the current from the charger will only be about 2 amperes.

Voltage Control

Many types of ac powered chargers designed to fill lead-acid batteries use voltage regulation to control the charge process. In some cases, the voltage is limited through the use of an electronic circuit. In other cases, the voltage is somewhat limited by the current capacity of the transformer used in the charger. Another method of voltage limitation is the built-in peak voltage output of the transformer, i.e. the voltage step-down ratio within the transformer.

It is common practice for charger manufacturers to set the voltage limit for a 12-volt battery at around 13.4 volts. At this limit it is unlikely that the battery, regardless of size, will be overcharged even if it is left on the charger for days. This is fine if the battery is to be filled over a period of several days. If the battery is to be filled in a day or less, then the voltage limit must be raised to over 15 volts. If ac power is to be used to charge large capacity lead-acid cells in deep cycle service, then the charger must have adjustable voltage output. This feature is not common in consumer car battery chargers.

If a 12-volt lead-acid battery is to be equalized, then the voltage output of the charger must be able to exceed 16 volts. A charge rate of C/20 is desirable during the equalizing charge. Chapter 2 details the parameters of charging lead-acid cells.

An Ac Powered Charger for Alternative Energy Service

When most people move to a place where commercial ac power is unavailable, their first purchase is usually a motor driven ac alternator. This unit supplies the household with the same type of power that the power company distributes, 120-volt ac power. Its big disadvantage, other than expense, is that there is no power unless the unit is running. Most users of ac alternators have a battery to provide low voltage (usually 12 volts) dc for lighting and whatnot when the ac alternator is turned off.

An ac powered charger is used to convert the higher voltage ac output of the motor driven alternator into low voltage dc to charge the batteries. If the battery pack has a capacity greater than 300 ampere-hours, the charger must be quite large and will be expensive. The charger must be capable of charging the battery at at least a C/10 rate. It is not cost effective to run a several kilowatt generator for long periods of time just to supply the battery with a few watts of charging energy.

Some commercially made battery chargers are set up for "Deep Cycle" service. These provide a high constant current to the battery. The voltage ceiling on these units is set very high, about 16 volts for 12-volt batteries. If a commercially produced charger is being considered it should be of this type. A simple and reliable high output charger can be homemade at about 1/4 the cost of these commercial units.

The heart of an ac powered charger is the transformer. Homebuilders should consider transformers which can handle more than 200 watts. The secondary of the transformer should be able to deliver at least 18 volts at an output current of at least 10 amperes. Transformers from large color televisions, electric welders, and custom made transformers are some sources of supply. The secondary can be modified by adding taps to it at various voltage levels. These taps will control the power output of the charger by limiting its voltage. These taps should be between 15 and 18 volts for a 12-volt battery system.

In any charger being powered by a motor driven alternator, full wave rectification of the ac is highly desirable. Half wave rectification wastes energy and increases the time that the motorized ac powerplant must be run to charge the battery.

If the secondary of the transformer has multiple taps in the proper ranges, the ac charger's output power can be controlled by selecting the appropriate tap. Metering is essential to ensure that the output of the charger is within the proper levels. Both output voltage and amperage should be accurately measured.

Figure 6-1 gives a schematic of an ac powered battery charger. This charger is designed to be simple and easy to build. It is not voltage regulated. If the proper transformer is used it will be capable of equalizing charges. If a transformer is not available with multiple taps on the secondary windings, one with only one output voltage may be used. Such a unit would not be adjustable. Make sure the diodes used in the rectifier have sufficient current handling capabilities to do the job. Heatsink these diodes well.

SOLAR CELLS

The photovoltaic, or solar cell converts sunlight directly into electricity. These units are simple to use, reliable, quiet, and very expensive. This section is not intended to be an exhaustive discussion of solar cells, but to provide some information on solar cells from the battery's point of view.

Solar Array Sizing

The most common error made in using solar cells to charge batteries is undersizing the output power of the solar array. Solar cells cost in the neighborhood of $10 per watt. At these prices it is easy not to buy enough of them to do the job. In most solar applications, the solar cells are not the only source of power. The solar system is usually

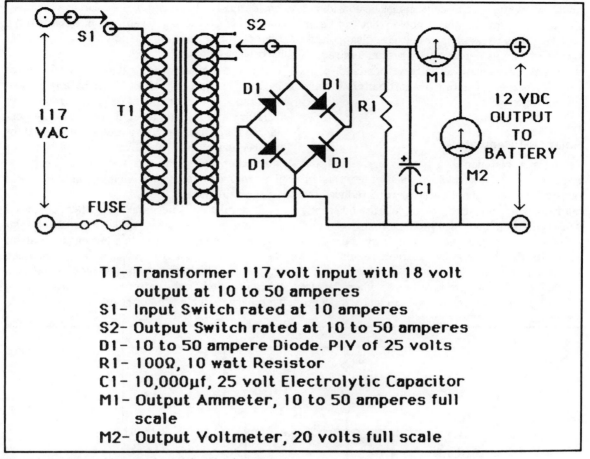

Fig. 6-1. Ac powered battery charger for 12-volt systems.

backed up by another power generating source.

The total accumulated power output of a solar array will vary greatly with a number of factors. Obviously, there are more sunny days in Florida than in Washington state. What we are concerned with here is the peak output power from the array to the battery. The sizing of the solar array to meet energy demands is a function of energy management and is discussed in Chapter 9.

The solar array should be able to charge the battery at a rate of C/20 or greater. For example, say we wish to charge a 12-volt 350 ampere-hour deep cycle lead-acid battery using solar cells. The output current from the panels to the battery must be in excess of 17 amperes. This amounts to an array output wattage of around 250 watts. At $10 per watt, this comes to about $2,500 worth of solar cells. This is the minimum necessary to charge a 12-volt, 350 ampere-hour battery.

This estimate merely considers the minimum energy necessary to effectively charge the battery. When we consider that energy is being used from the battery to power various appliances, the wattage output of the array must be larger. If we allow for cloudy days and the inefficiency of the cells during the early morning and late evening sun, the array must be larger still. In other words, the solar array will only provide its rated energy output about four hours per day and then only on sunny days.

Even in very small alternative energy systems,

about 500 watts of solar cells should be considered as the minimum amount of power that can be effective. Many users of solar cells are disappointed by their systems because they are undersized. In undersized systems, the battery spends all of its time discharged, resulting in short battery life and poor efficiency.

Solar Cell Regulators

It is a common practice to use an output regulator on solar arrays. In most cases, this expensive regulator is not necessary and reduced the entire system's efficiency. Solar cells are inherently self-regulating. They will produce their rated output and no more. As such, if the array is properly sized to the battery, no further regulation is necessary.

Solar cells only produce power when the sun is shining on them. If nights and cloudy days are considered, it takes a very large solar array to overcharge a large battery. Even very powerful arrays can be run without regulation if the battery is sized large enough.

Charge regulators are only necessary in arrays that are oversized in relation to the battery capacity. If the array is capable of delivering more than a C/10 rate of charge, a regulator should be considered if no energy is being used from the battery on a regular basis. If the array can deliver C/5 or more, then a regulator is necessary. Very few photovoltaic systems have enough output power to require regulation.

If regulation is necessary, Fig. 6-2 provides a schematic of an electronic regulator for 12-volt solar systems. This circuit will do a better job at about one-fifth the cost of commercially produced regulators. This regulator is a *shunt type* and uses energy only when the battery is in danger of being overcharged. This type of regulator is more effi-

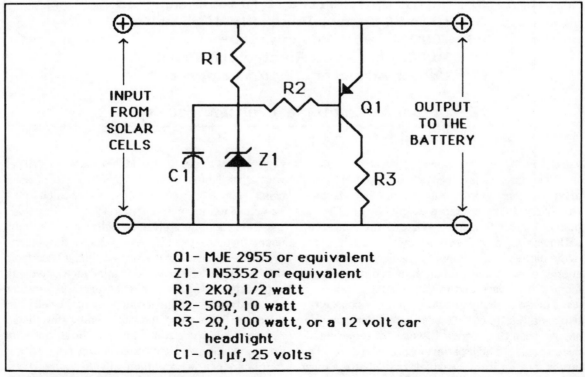

Fig. 6-2. Shunt regulator for solar cells.

cient than the *series type* which has losses even when it is not regulating. Changing the zener diode will make the regulator work on either 24-or 48-volt systems.

Equalizing Batteries in Solar Systems

One drawback of solar cells is the difficulty in equalizing the cells within the battery. Chapter 2 discusses the equalizing charge for lead-acid cells.

The equalizing charge is a controlled overcharge of already full batteries. This condition is difficult to produce with the limited energy available from solar cells. The array must be very large to function as an equalizing charge source for the batteries. In addition, the sun must be shining. It is highly desirable to have a second source of controllable energy for equalizing charges.

A MOTORIZED CHARGER FOR 12-VOLT SYSTEMS

In applications using alternative energy as a power source there are always periods of low energy production. There are times when the sun doesn't shine and the wind doesn't blow. During these periods the battery is not being charged; energy is being consumed from it. Solar and wind systems must have oversized capacity in the battery to compensate for these periods where no energy is being produced.

There is no way to avoid having to oversize the battery. A small motorized low voltage dc charger can supply energy for the periods of time when the solar or wind power is not happening. This type of charger can stand alone—it can be the sole power when it is time to run an equalizing charge on the battery.

This type of motorized charger is not commercially produced, but is simple enough for home construction. It has the added advantage of being very inexpensive. The cost of establishing and running this motorized charger is very low. Even considering the cost of engine maintenance and gasoline for a 10-year period, this system costs less than 30 percent of a solar or 50 percent of a wind system of the same output.

Lawnmower Engines and Automobile Alternators

The automotive alternator is a cheap and available low voltage dc power source. These units have amperage outputs from 35 to 200 amperes. Automotive alternators can produce high amperage dc charging current for the battery at a miniscule expense compared to ac powered chargers, wind, or solar sources.

In order to produce energy the alternator must be turned. In an automobile this is done by the same engine which moves the car. A lawnmower engine can be used to drive the automotive alternator to its full rated output. Horizontal shaft gasoline motors rated at 3.5 to 8 horsepower will work best. The horsepower output of the engine should match the amperage output of the alternator. A 3.5 horsepower engine will drive a 35 ampere alternator. If the alternator is 50 to 75 amperes, then 5 horsepower is required. An 8 horsepower motor is required to drive alternators with outputs over 125 amperes.

In fact, any power source can be used to drive the alternator. Those with commercial ac power can use an electric motor. If wind or water power is available it can drive the alternator. The small gasoline engine is used here as an example because it is very cheap, available, powerful and allows the user to have power when and where it is needed.

The mechanical construction is simple. Figure 6-3 shows a diagram for harnessing a lawnmower engine to an automotive alternator. Both the motor and the alternator are bolted to a base plate. The base plate can be a thick wooden plank or a piece of steel plate. The motor is connected to the alternator with a pulley and a "Vee" belt. Care must be taken in the shaft and pulley alignment. A five- or six-inch diameter pulley works well on the motor. The stock pulley is fine on the alternator; it will accept a standard 1/2-inch vee belt. If the stock mounting bracket is available for the alternator it can be used. If not, a bracket can be made from two pieces of steel angle stock. A turnbuckle works well as a tensioner for the belt.

The small one cylinder gasoline engine is universally available. The cost of a new 3.5

Fig. 6-3. Motorized dc charger using a 12-volt car alternator.

horsepower unit is about $130. Very well made engines around 5 horsepower are about $350. Small gas motors are very commonly used. Old lawnmowers are an excellent source.

A motorized charger of this sort can charge and equalize even very large batteries. If all of the parts for the charger are purchased new, the cost will be under $500. If recycled parts are used, the unit can cost as little as $200. With outputs over one kilowatt, these units are the cheapest form of high amperage, low voltage power available. They can provide battery charging energy for places which cannot be reached with commercial power.

Automotive Alternators

Alternators are mass produced by the millions to charge the battery and provide power to the automobile. There is one riding under the hood of every car on the road. There are millions of them languishing in junk yards. They can be purchased new, rebuilt, or from junk yards. Costs vary widely with the output power of the alternator and place of purchase. A rebuilt 35-ampere model costs about $40. The rebuilt 100-ampere type costs about $150. Prices are much lower on used units.

The automotive alternator generates electricity by rotating a magnetic field within a circle of stationary copper wire windings. The magnetic field is powered by electricity from the battery. This rotating electromagnet is known as the *rotor*. It is attached to the alternator's main shaft, which holds the pulley. The stationary copper windings are known as the *stator*. The stator is housed in the alternator case. There are also diodes heatsunk to the alternator case. These diodes rectify the output of the alternator to direct current for the automobile and the battery.

When the rotor is rapidly turned, a current of about 1 ampere to the electromagnet on the spinning rotor will induce 30 to 50 amperes of current in the stator. Alternators are mechanically powered current amplifiers. Internally the alternator is producing 3-phase ac current which is rectified by the diodes in the alternator case into dc current for the battery. The alternator is designed to reach its rated power when rotated at between 3,000 and 6,000 revolutions per minute.

Automotive alternators vary widely in power output and physical construction. In general, the higher the amperage output of the alternator the more efficient it is. The higher output alternators are larger in diameter and have larger size wire in the stator. The alternator should be sized so it can deliver at least a C/20 rate of charge to the battery. Some types of alternators can be changed from low ampere models to high ampere models by simply changing the stator.

The automotive alternator is an incredibly rugged device. It is designed to generate its full rated output power under the hood of a running car on a hot summer's day. These units seem impossible to overheat when running outside on a motorized charger. Other than the maintenance of brush and bearing replacement they are virtually trouble free when used in alternative energy service.

Automotive Alternator Control Systems

The motorized alternator requires a control system to prevent overcharging or too rapid charging of the battery. In the automobile this task is accomplished by the voltage regulator.

The automotive voltage regulator is designed to keep the entire system voltage at about 13.8 volts. If the system voltage is below 13.8 volts, the regulator causes the alternator to put its entire rated output power into the system until the voltage does reach 13.8 volts. This entire alternator output is often too much for the fully discharged deep cycle battery. It will be charged too quickly.

Automotive voltage regulators will not work in deep cycle service. The automotive voltage regulator will attempt to fill a fully discharged battery at rates which are much too fast. In automotive service, the battery is usually discharged less than 1 percent of its capacity. The automotive control system is designed for this very shallow type of float service. In deep cycle service, the battery should be filled at a controllable rate (not greater than C/10 or less than C/20) until it is full. Then voltage regulation is acceptable.

The limit at which the voltage is regulated is not adjustable in most automotive voltage regulators. In a float type service such as in an automobile, the voltage limit of 13.8 volts works well. In deep cycle service, the limit must be raised to over 15 volts in order to completely fill the empty battery in a short period of time. Adjustability is also required for equalizing charges.

The simplest form of control is to insert a variable resistor (a rheostat) between the alternator's rotor input and the positive pole of the battery. A rheostat of 25 Ω rated at 25 watts works well. Figure 6-4 shows a schematic for wiring a rheostat into the alternator/battery system.

The rheostat will control the amount of energy fed to the alternator's rotating electromagnet, and thereby the alternator's output current. The only problem with this system is that while the output amperage of the alternator is controlled, the output voltage is not. This type of control turns the alternator into a constant current source. The system voltage will vary depending on the state of charge of the battery.

The absence of voltage regulation makes it necessary to monitor the system when it is charging the battery. The battery can be overcharged if left on the charger for too long a period of time. The voltage in the system may rise to over 16 volts and damage equipment left on line during the charge. What is needed is both amperage and voltage control of the system.

The Electron Connection Ltd., P.O. Box 442, Medford, Oregon 97501 manufactures a control system for automotive alternators which will regulate both amperage and voltage. This control system is totally electronic and uses the highly efficient pulse width modulator (PWM) to control the alternator's field rotor. By using this control system, the battery can be filled at a constant amperage rate until it is fully charged. When the battery is full, then the voltage control section takes over. This control system is user adjustable to fit all sizes of alternators and battery packs.

WIND AND WATER POWER SOURCES

Wind machines are becoming more common as a source of energy. Modern materials and engineering techniques have made wind power practical in small applications. In general, wind machines are cheaper per kilowatt than are solar cells.

If a wind machine is used to charge the battery it must be capable of at least a C/20 rate. If the wind machine cannot do this, then it is undersized in relation to the battery. As with the solar system, it is necessary in most wind applications to have a second power source available.

Wind machines can sometimes provide large amounts of power continuously for several days at a time. Regulation of the charge energy is necessary to prevent overcharging the battery. Straight voltage regulation of the entire system is sufficient.

Water powered sources are interesting from the battery's point of view. Many water units are capable of continuous operation and as such batteries are not strictly necessary. If the water source generates enough power to supply all needs, then battery storage is not required. In systems where the generation capacity is not enough to meet peak needs, it is common to store energy in batteries.

Batteries used in water systems should be considered as used in float type service. As such, the system must be voltage controlled. The voltage control point should be located at the batteries. This is done to minimize inaccuracy due to voltage losses through the wiring.

If there is sufficient water power to meet all demands, and this power is continuous, then batteries are not needed for storage. Regular 120-volt ac housepower is the type to generate and use. Low voltage dc systems are not compatible with the long lengths of wiring common to water systems. In stand-alone water systems it is cheaper and more efficient to maintain a motorized ac alternator for backup power rather than to use batteries.

SMALL NI-CAD CHARGERS

The average small portable device comes with a charger to fill the batteries. These constant current chargers require that the user keep track of

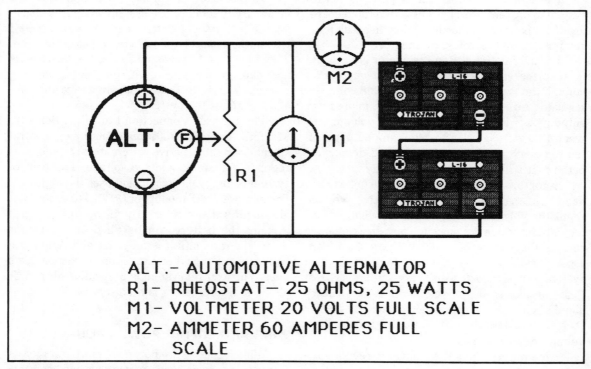

Fig. 6-4. Diagram for a motorized charger—12 volts.

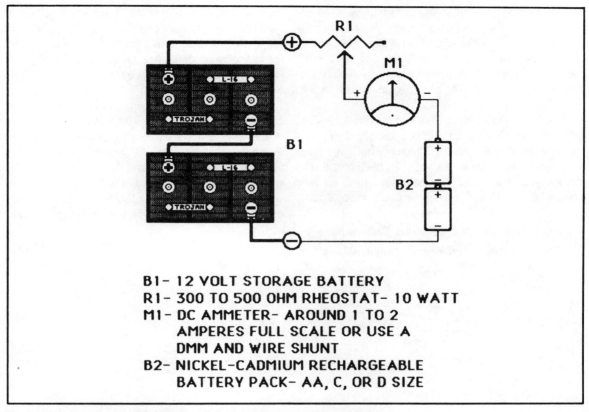

B1– 12 VOLT STORAGE BATTERY
R1– 300 TO 500 OHM RHEOSTAT– 10 WATT
M1– DC AMMETER– AROUND 1 TO 2
      AMPERES FULL SCALE OR USE A
      DMM AND WIRE SHUNT
B2– NICKEL-CADMIUM RECHARGEABLE
      BATTERY PACK– AA, C, OR D SIZE

Fig. 6-5. Constant current charging of small ni-cads using a 12 volt battery.

the amount of time the battery spends being charged. This is very important if the constant current method of charging is used on any type of battery regardless of capacity. Read the manufacturer's directions for recharging the battery and follow them.

A Battery Powered Small Ni-Cad Charger

Users of battery based alternative energy systems cannot use the ac chargers supplied with most of the appliances. The small ni-cads can be charged from the main storage battery. Figure 6-5 gives a schematic for using a large 12-volt lead-acid battery to charge small nickel-cadmium cells.

The ni-cad is charged using a resistor to control the charging current. The value of this resistor depends on the voltage of the ni-cad battery pack and on the charging current required by the ni-cad cells. For example, consider a ni-cad battery composed of two AA cells in series. The ni-cad pack has a voltage of about 2.5 volts and a charging current of 0.05 amperes (50 milliamperes) for 12 to 14 hours. Subtract the voltage of the ni-cad pack from the voltage of the main storage battery. In this case, 12 volts minus 2.5 volts equals 9.5 volts. Using Ohm's Law in the form $R = E/I$ we can compute the amount of resistance needed to limit the current to whatever amount is needed. In this case, R equals 9.5 volts/0.05 amperes, or R equals 190Ω. The amount of power dissipated by a resistor is expressed as $P = I^2R$. In this case, P equals 0.05 amperes times 0.05 amperes time 190 ohms or P = .475 watts. A 1/2 watt resistor will do nicely. The resistor does not have to be the exact value computed. Anything that is within 20 percent of the computed value will work well.

A variable resistor (rheostat) can be used in place of the fixed value resistor. A rheostat of around 300Ω to 500Ω at 5 to 10 watts will handle most types and sizes of small rechargeable batteries. If a rheostat is used, be sure to include an ammeter to measure the charging current flowing into the battery.

Some manufacturers of ni-cad powered devices offer a *car charger* to fill the device from an automotive 12-volt system. These are nothing more than a resistor in series between the ni-cads and the car system. These car chargers usually sell for between $8 and $20. The resistor costs about $ .20.

Commercial Nickel-Cadmium Chargers

Most commercial chargers for small ni-cads are the constant current variety. They are available at moderate costs, between $20 and $60. Some of these machines contain a timer which will shut off the charging process after a set period of time. This is a very good feature. It prevents the ni-cads from being overcharged.

Another type of commercially made charger contains both amperage and voltage regulation. With this type of power source, the small ni-cad battery can be charged rapidly and still left charging for extended periods of time. Batteries charged by such a machine are always full when they are removed from the charger. This type of charger is usually programmable to fit any capacity and voltage ni-cad battery pack.

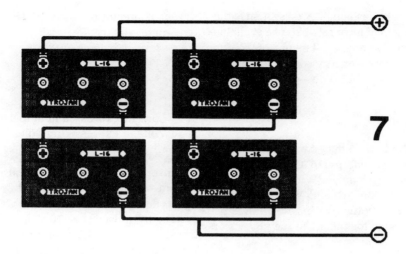

Using Batteries Effectively

If a battery is to provide satisfactory service it must be used efficiently within the system. An electrical circuit is like a chain. It is limited by its weakest part. No matter how much battery capacity is available, it cannot be effective if it is improperly used. Proper application is not difficult, but attention to many details is essential.

Details such as wiring size, length of wiring, switches, and the manner in which everything is interconnected are of critical importance in low voltage systems. In a low voltage system there is much less voltage to waste in wiring and switching losses. If a battery application is to be effective, all portions of the circuit must have very low resistance paths.

The information presented in this chapter is directed towards users of home low voltage (12 to 48 volts) energy systems. These applications use the low voltage dc energy directly from the battery. The information here is applicable to anyone using low voltage dc for power, whether stored in a battery or not, such as in houses using dc for appliances, recreational vehicles, trucks, and boats.

Those considering alternative energy for home power should examine the existing wiring within the home. It is quite possible that it is not suitable for low voltage service. Very few homes are wired for low voltage dc. The wiring within the walls is usually too small and poorly interconnected for dc service. The existing ac wiring will make a very poor path for low voltage dc power. Too much energy will be lost within the wiring; the system will be very inefficient.

Retrofitting an old house with a low voltage dc system can be very expensive. The cost of commercial rewiring often exceeds the cost of the alternative energy equipment. The installation of an inverter can make it possible to use the existing standard house-wiring efficiently. Information on inverters is contained in Chapter 8.

If new construction is being planned, the installation of low voltage dc wiring should be considered. If the house is properly wired, it is more efficient to use the dc energy directly from the battery than in any other fashion. The dc system is the cheapest, simplest, and most reliable type available to users of alternative energy. The use of 12 Vdc is common in alternative energy households due to the high availability of 12 Vdc lighting and appliances.

WIRING

A battery stores the energy within a system. Wire transports this energy to the various appliances where it is consumed. Wiring techniques which work well in standard high voltage applications (120 Vac) are ineffective in low voltage dc systems. The reasons for this lie in the physics of electricity.

Ohm's Law and Wiring Resistance

Every piece of wire, every connection, and every switch within the system has some resistance. The resistance of a piece of wire depends on its composition (copper or aluminum), its diameter (gauge size), and its length. The total resistance within a circuit greatly affects the efficiency of a low voltage system. If the total resistance is too great, so much energy will be lost that the appliances will not function.

The relationship between voltage, current, and resistance is best expressed on Ohm's Law:

$$E = IR$$

E = Voltage expressed in volts
I = Current expressed in amperes
R = Resistance expressed in Ohms (Ω)

The power lost in a piece of wire is apparent as a voltage loss across the wire. Since the resistance of the wire is a constant, the voltage drop across the wire is directly proportional to the amount of current flowing through the wire. The more current flowing through a wire, the greater the voltage loss within the wire.

A 12-volt system must move electrons 10 times faster than a 120-volt system in order to transfer the same amount of power. Power is equal to current times voltage:

$$P = IE$$

P = Power expressed in watts
I = Current expresses in amperes
E = Voltage expressed in volts

In order to transfer 120 watts in a 120-volt system we must move a current of 1 ampere (120 watts equals 1 ampere times 120 volts). In a 12-volt system we must move 10 amperes of current to transfer the same 120 watts of power (120 watts equals 10 amperes times 12 volts). Since the 12-volt system has 1/10 the voltage of the 120-volt system, ten times as much current must be moved to transfer the same amount of power. In household low voltage service, this means very low resistance in the wiring is essential due to the large amount of current required to transfer any given amount of power.

Ohm's Law tells us that the voltage drop across a wire is directly proportional to the amount of current flowing through the wire (E = IR). For example, consider a length of wire with a resistance of 1Ω. In order to transport 120 watts of energy through this wire at 120 volts, we must move a current of 1 ampere. The voltage drop in the wire is equal to the current times the resistance of the wire: E = IR. In the case of 120 volts, the loss is equal to 1 ampere times 1 Ω, which is 1 volt (1 volt = 1 ampere $\times$ 1 Ω). In a 12 volt system we must move 10 amperes of current to transport the same 120 watts of power. Here the voltage loss is again equal to the product of the current and the wire's resistance: voltage drop = 10 amperes times 1 Ω, which is 10 volts (10 volts = 10 amperes $\times$ 1 Ω). The voltage loss through the same wire is ten times greater on 12 volts than it is on 120 volts.

In the 120-volt system, if 1 volt is lost, this is not very much. It is less than 1 percent of the energy being transferred. In the 12-volt system, 10 volts is lost and that is over 83 percent of the energy. A 120-watt device which requires 12 volts for power would only get 2 volts at the end of this 1 Ω wire. It would not come close to operating.

Obviously, if low voltage electricity is to be used for power then the wiring must be of low enough resistance. Only by minimizing the amount of resistance in the complete circuit can low voltage electricity be efficiently transferred.

In order to find out how much resistance a piece of wire has, it is necessary to consult a Copper Wire Table. Copper wire is the only material to use in low voltage applications. Aluminum wire has more resistance and is very difficult to connect in a low resistance fashion.

The Copper Wire Table

Table 7-1 is a Copper Wire Table. This table provides a variety of information on various sizes of annealed solid copper wire. Wire is sized by gauge number. The larger the gauge number the smaller the diameter of the wire. The larger the gauge number the more resistance the wire has per unit of length.

The resistance information in the table is given in ohms per 1000 feet and in ohms per kilometer. To find the resistance of a single foot, divide the ohms per thousand feet figure by 1000. Resistance is also given in the table in feet per ohm and meters per ohm. The reciprocal of feet/ohm is ohms/foot, which is also a way to find ohms per foot by using a calculator.

There is other information included in the table. The diameters of various gauges of wire is expressed in mils (thousandths of an inch) and in millimeters. This information is computed at a temperature of 68° F. (20° C.).

How to Calculate Wiring Loss

With the Copper Wire Table it is possible to find the total resistance of a length of wire. It is necessary to know the gauge size of the wire and its length. Multiply the length of the wire in feet by its resistance per foot. The resulting product is the total resistance for that particular gauge of copper wire at that particular length.

For example, say we have 80 feet of 12-gauge copper wire. The Copper Wire Table gives us a resistance of 1.588 Ω per 1000 feet for 12-gauge wire. When divided by 1000 this resistance figure is 0.001588 Ω per foot. Therefore, 12-gauge copper wire has a resistance of 0.001588 Ω per foot. If we are using 80 feet of this wire, then the total resistance of the entire length of wire is equal to 0.001588 Ω per foot times 80 feet, which is 0.127 Ω.

R = (Ω/ft.) (ft.)
R = (0.001588 Ω/ft.) (80 ft.)
R = 0.127 Ω

So the entire 80-foot length of 12-gauge copper wire has a total resistance of 0.127 ohms.

This 80-foot length of wire is capable of powering an appliance that is 40 feet from the power source. A complete circuit is necessary—40 feet to the appliance and 40 feet back again. It is a two-wire circuit; a return path for the current is necessary in order to complete the circuit.

Now assume that we are to wire up a 40-watt light bulb using this 80 foot length of 12-gauge wire. Also assume that the system voltage is 12 volts. At 12 volts, a 40 watt light bulb will draw 3.33 amperes.

P = IE

$$I = \frac{P}{E}$$

$$I = \frac{40 \text{ watts}}{12 \text{ volts}}$$

I = 3.33 amperes

The voltage loss in the wiring is equal to the total resistance of the wire times the amount of current flowing through the wire. In this case, the voltage loss is equal to 3.33 amperes times 0.127 ohms which is 0.423 volts.

E = IR
E = (3.33 amperes) (0.127 Ω)
E = 0.423 volts

Table 7-1. The Copper Wire Table.

WIRE GA.	RESISTANCE				DIAMETER	
	OHMS PER 1000 FEET	FEET PER OHM	OHMS PER KILOMETER	METERS PER OHM	DIAMETER IN MILS	DIAMETER IN MM.
0000	0.04901	20400	0.1608	6219	460.0	11.68
000	0.06180	16180	0.2028	4932	409.6	10.40
00	0.07793	12830	0.2557	3911	364.8	9.266
0	0.09827	10180	0.3224	3102	324.9	8.252
2	0.1563	6400	0.5127	1951	257.6	6.544
4	0.2485	4025	0.8152	1227	204.3	5.189
6	0.3951	2531	1.296	771.5	163.0	4.115
8	0.6282	1592	2.061	485.2	128.5	3.264
10	0.9989	1001	3.277	305.1	101.9	2.588
12	1.588	629.6	5.211	191.9	80.81	2.053
14	2.525	396.0	8.285	120.7	64.08	1.628
16	4.016	249.0	13.17	75.90	50.82	1.291
18	6.385	156.6	20.95	47.74	40.30	1.024
20	10.15	98.50	33.31	30.02	31.96	0.8118
22	16.14	61.95	52.96	18.88	25.35	0.6438
24	25.67	38.96	84.21	11.87	20.10	0.5106
26	40.81	24.50	133.9	7.468	15.94	0.4049
28	64.90	15.41	212.9	4.697	12.64	0.3211

The voltage at the light bulb is 11.577 volts (12 volts − 0.423 volts = 11.577 volts); enough to operate it. The amount of power lost in the wiring can be computed by dividing the voltage drop in the wiring by the input voltage. In this case power loss (in percent) is equal to 0.423 volts divided by 12 volts times 100 percent, which is 3.5 percent.

$$\text{Power Loss in \%} = \frac{\text{Voltage drop}}{\text{Voltage In}} (100\%)$$

$$P = \frac{0.423 \text{ volts}}{12 \text{ volts}} (100\%)$$

$$P = 3.5\%$$

In general most 12-volt lighting and appliances will work on voltages as low as 11 volts. Wiring should be sized so that at least 11 volts is available to the appliance. Power losses no greater than nine percent are acceptable in 12-volt systems; a nine percent loss is about one volt in a 12-volt system. If possible the losses should be less than five percent which is about a 0.5 volt drop.

Consider once again the 80-foot length of 12-gauge wire. In this example let's assume we wish to locate a wind powered generator 40 feet from the house, where the battery is located. This wind machine is capable of delivering 20 amperes to the battery. The voltage drop in the 80-foot complete circuit of 12-gauge wire carrying 20 amperes is 2.54 volts. Twenty-one percent of the energy being transferred is lost within the wiring. This is obviously not acceptable.

The amount of energy lost within the wiring is determined by three factors: the length of wire in the complete circuit, the resistance of the wire, and the amount of current flowing through the wire. The more current the wire must carry, the larger in diameter it should be. This translates to a lower gauge number. A quick glance at the Copper Wire Table shows that 10-gauge wire has less resistance per 1000 feet than does 12 gauge. The higher the current in a circuit, the lower the gauge number of the wire which should be used.

The longer the circuit is, the larger in diameter the wire should be. The effect of the resistance is cumulative. For the same gauge wire, twice the length of wire will have twice the resistance. Long runs of wiring should be made with wire of low gauge number (large diameter).

Techniques for Specifying Wire Size

The information for computing wiring loss can be used to find the correct gauge of wire to handle any particular job. The information necessary to do this concerns the wattage or amperage of the appliance, its distance from the battery, and the system's voltage.

Specifying wire size is the process of finding the wire loss in reverse. We know the wattage of the appliance, the maximum voltage drop, the distance from the battery, and the system's voltage. We need to know which wire gauge size to use to get efficient operation.

Here is an equation which provides the resistance per 1000 ft of wire that will meet certain requirements.

$$R = \frac{E}{I\,L}(1000)$$

R = Resistance expressed in Ω per 1000 feet
E = Maximum allowable voltage drop in the wiring expressed in volts
I = Amount of current flow through the wire expressed in amperes
L = The length of the wire in the complete circuit expressed in feet

Note that the length of the wiring is measured for the entire circuit. If a device is 100 feet from the battery it will take 200 feet of wire to make a complete two-wire circuit.

This equation gives a resistance value expressed in ohms per 1000 feet as an answer. Simply locate the wire size on the Copper Wire Table that has that amount of resistance or less per 1000 feet.

For example, assume that we wish to locate a device 20 feet from a 12-volt battery. This device will draw 8 amperes of current. The maximum allowable voltage drop through the wiring is 1 volt.

$$R = \frac{E}{I\,L}(1000)$$

$$R = \frac{1 \text{ volt}}{(8 \text{ amperes})(40 \text{ feet})}(1000)$$

$$R = 3.125 \text{ Ohms per thousand feet}$$

The resistance of the wire needed must be less

than 3.125 ohms per thousand feet. The Copper Wire Table tells us that the nearest wire gauge size which has less than 3.125 ohms per 1000 feet is 14-gauge wire. Fourteen gauge wire or bigger will meet the demands of this circuit.

The fact that 14-gauge wire will work in the above example does not mean that we shouldn't use wire that is larger than 14 gauge. In low voltage systems using batteries, the name of the game is efficiency. If we were to use 12- or 10-gauge wire in the above example the losses would be even less. The formula given allows the user to input the maximum allowable voltage drop. In a 12-volt system this is 1 volt. However, a drop of 0.5 volts is better. The system cannot be too efficient.

Low Voltage Wiring Techniques

In reality, houses contain many circuits. Some of these circuits are wired in parallel, i.e. two or more devices may be supplied power by the same piece of wire. There are general low voltage wiring techniques which will meet most common residential requirements.

The battery may be connected to two large diameter (low gauge number) wires that run the length of the building. Such a heavy pair of wires is known as a *bus*. Each light or outlet is attached directly to the bus with smaller diameter wires. This structure is similar to the skeleton of a fish; a heavy spine with smaller bones attached to it.

Buses should be made from wire sized 6 gauge or larger for runs less than 100 feet. For wiring lengths over 100 feet, use 4- or 2-gauge wire. The branch wiring can be of smaller size—10 or 12 gauge. It is important that each branch wire supply only one or two devices. Ideally, each appliance should have its own branch circuit to the main bus. The branch circuits should be sized with allowances being made for the length of the branch and the current to be transferred by the branch. Each branch circuit should be soldered to the bus.

Heavy copper wire is an expensive item. It is often cheaper to parallel several smaller wires to get a lower resistance conductor than it is to purchase large diameter wire. Take for example the common Romex cable used in wiring houses (type NM 12-2). This cable contains three wires; each of which is 12 gauge. Two of the wires are insulated within the cable and one is not. This cable is intended for use in house wiring at 120 Vac. It is relatively inexpensive as it is mass produced.

In low voltage applications the entire cable can be used as a single conductor by paralleling all the wires within it. As such, two separate pieces of cable are necessary to provide a complete circuit. Each three wire cable will electrically become a single wire with a resistance one-third that of a 12-gauge wire. Twelve gauge wire has a resistance of 1.588 Ω per thousand feet. By paralleling three 12-gauge wires, the resulting resistance is 0.529 Ω per thousand feet. This is roughly equivalent to a copper wire of 8 gauge. The NM 12-2 cable is often much cheaper than the same length of 8 gauge wire.

The NM 12-2 cable can also be used in the low voltage branch wiring. In this case, there is an extra conductor within the cable; the uninsulated grounding wire. This extra wire is necessary in the three-wire scheme used in house 120 Vac wiring. In low voltage systems only two conductors are needed while there are three conductors in the cable. The uninsulated grounding wire can be paralleled with either of the two other insulated wires. This will reduce the entire circuits resistance.

CONNECTIONS, SWITCHES, FUSES, AND OUTLETS

Every component in a series circuit is of critical importance. Even the largest wire will not perform well if it is not connected in a low resistance fashion. Each switch can become a bottleneck for the electrical flow if its contact resistance is excessive. The connections in and out of fuses are another source of cumulative resistance.

It is not enough to have a large battery, wired up with large wire. Every connection and component in the circuit must have its resistance minimized. It only takes one point of high resistance within the circuit to limit the power flowing through the entire circuit.

Soldered Connections

Soldered connections should be used in low voltage systems wherever it is possible. All wire-to-wire connections should be soldered. Keep all mechanical connections to an absolute minimum. The practice of twisting the wires together and covering them with a wire nut is common in 120 Vac housewiring. This mechanical technique is not acceptable in low voltage systems. Over a period of time the wire will oxidize and the mechanical connection's resistance will increase.

Soldering a connection makes a permanent low resistance joint. The process of soldering is not difficult as long as some basic rules are followed. All the wires to be soldered must be clean and bright. Use a high grade rosin core solder with a good flux, such as Kester 44. Melt the solder on the work, not on the iron. Be sure to use an iron of sufficient wattage to do the job.

Having sufficient wattage to solder several 12-gauge wires to a 4-gauge wire is a problem. These wires are very massive and carry heat away from the joint quickly. Even the largest 260-watt commercial electric soldering guns do not put out heat fast enough to do this job. One must resort to using a propane torch to heat the wire for soldering. Be careful not to oxidize the wire with the torch, as this will result in a poor joint of higher resistance. Making a good soldered joint takes practice. Keep at it until you get it right.

Mechanical Connections

There are always some mechanical connections within a system. Major system components such as the battery and the power source are usually mechanically connected. It is important that these mechanical connections have as little resistance as possible. One bad mechanical connection can make the entire circuit inoperative.

Some sort of connector must be placed on the end of the wire. The common commercially available ring connector is acceptable for currents less than 10 amperes. The connectors must be soldered after crimping. The mechanical crimp connection between the wire and the connector will gradually oxidize and become resistive if it is not soldered.

The mechanical connections to the battery, the bus, and to an inverter must be able to handle large amounts of current. Currents over 100 amperes through the main bus are common in alternative energy systems. this requires big wire—from 4 to 0000 gauge depending on the situation. The cables necessary for battery interconnect, bus interconnect, and from the batteries to the inverter are not available commercially. Commercial cables are not soldered and will be attacked by battery acid over a period of time. These cables must be capable of surges in the neighborhood of 200 amps. I recommend you make your own.

The cables from the batteries to the inverter should be at least 0-gauge copper wire, while the battery interconnect cables should be 4-gauge copper or larger. This wire is expensive, so keep all the runs as short as possible. Also short runs have less electrical loss.

To make ends on these cables do as follows:
1. Strip the insulation from 2.5 inches of the cable end.
2. Take the twist out of the cable so the strands run roughly parallel and give them a very light coat of paste soldering flux.
3. Insert the stripped cable end into a piece of clean copper tubing 3.5 inches in length. Be sure the interior of the tubing is clean and bright. If it is not, then polish it with steel wool. Use 1/2 copper tubing for 0 gauge and 3/8 inch tubing for 2 or 4 gauge.
4. Squash the tubing in a vise so the entire assembly is flattened. Use a hammer to flatten the entire assembly if a vise is not available.
5. Fold over the 1-inch section of tubing that does not have wire within. This is done so that the solder will not run out. Fold over 1/2 inch from the end.
6. Clamp the wire in a vise, with the connector dangling below the vise. Heat the bottom folded portion of the connector with a propane torch. Flow solder down into the tubing. Fill the tubing with solder. Use a good grade solder—Kester Radio TV solder 60 percent tin, 40 percent lead is recommended.

7. Locate and drill the appropriate hole through the connector.

8. Use a hacksaw to saw off the folded end of the connector that does not contain any wire.

9. Deburr and smooth the connector with a file. Polish it bright with some sandpaper. Make sure the connector is flat. If it is not, then simply flatten it with a hammer.

This technique will result in a massive, long lasting, low resistance connector that is far better than anything which can be purchased commercially. They are a lot of trouble to make but well worth the effort.

No matter what type of mechanical connection is made, it is not permanent. Metals oxidize, and their oxides are very poor conductors of electricity. All mechanical connections will have to be periodically disassembled and cleaned. This is especially true of the connections to the batteries themselves. The presence of acid or basic chemicals within the cells will greatly increase the rate of corrosion of the battery's connectors. Simply disassemble the connections, polish them with steel wool until they are bright and shining, and then reassemble them.

Switches

The switches in a low voltage system must have low losses just like the rest of the system's components. Commercial switches which are designed for home service on 120 Vac will work if they are not heavily loaded. In 120 Vac service a switch may be rated to handle 15 amperes. This same switch can be used in a low voltage system if the current does not exceed five amperes.

The reason for derating 120 Vac switches to one-third in low voltage service is loss. The contact resistance of the switch limits the current flow through the switch at about 5 amperes. At amperages over 5 amperes, the switch will have too much voltage drop across it. Significant amounts of power are wasted within the switch in the form of heat.

Automotive switches often have less loss than standard wall switches. The automotive switch is designed to operate at low voltages. It has less contact resistance and is more efficient. They are commonly available in current ratings as high as 20 amperes.

There is a problem in switching currents over 50 amperes. Commercial switches are not commonly available in high amperage ratings. Where they are available, they are very expensive. The use of high amperage switching on low voltage systems is discouraged except where it is absolutely necessary.

One answer to high current switching is the solenoid. A solenoid is an electrically operated switch. Solenoids are available that can carry several hundred amperes of current. The solenoid should be capable of continuous operation (most are not). One major disadvantage to the solenoid is that it takes power to run it. Another is price; they are expensive.

Large open knife switches were in common use once. If one of these can be located, it may be used in a low voltage system. These large knife switches are ideal for high current, low voltage use. They are massive and have very low loss. They have fallen into disuse in common wiring because they are exposed and present a shock hazard on higher voltages.

No matter what type of switching is used, the connections to and from the switch must be well made. Solder the wiring to the switch if it is possible. If the connection must be mechanical, make sure the parts are clean and bright. Make sure all the mechanical connections are very tight. High contact pressure assures a lower resistance connection.

Fuses and Circuit Breakers

The mechanical connections in a fuse panel are a potential source of loss. These connections should be periodically cleaned. The standards screw-in house fuse is preferable to the small automotive type. The house fuse has much more contact area. Automotive cartridge-type fuses should not be used because of their small contact area.

Size the fuses in the range of 30 amperes to 60

amperes. A direct short circuit on a properly wired low voltage system will transfer over 100 amperes. As such, a 30- to 50-ampere fuse is adequate short circuit protection even though an appliance may only draw a few amperes. Fuses are actually resistors. The higher the amperage rating of the fuse the lower its resistance. It is not efficient to introduce any more resistance into a low voltage circuit than is absolutely necessary.

Circuit breakers have less loss than fuses in most cases. This is only true of breakers which are designed for low voltage operation. Standard 120 Vac circuit breakers are not suitable for low voltage use. Fortunately, automobiles, boats and recreational vehicles are low voltage systems. Circuit breakers designed for automobile or boat service are ideal for usage in alternative energy systems.

Again, be sure all mechanical connections to and from the fuses or circuit breakers are well made. Solder the wiring wherever possible.

Outlets and Plugs

If a light or an appliance can be soldered directly to the wiring, this should be done. In some cases however, we may wish to have an outlet into which we can plug a low voltage appliance. Both the socket and the plug are part of the complete circuit. If either has high resistance, then it will be the limiting factor in the circuit.

Standard outlets and plugs for use in houses at 120 Vac are usually rated at 15 amperes. They may be used for current flow of not greater than 5 amperes in low voltage systems. Be sure the mechanical connections to the plug or socket are well made.

Care must be taken in the polarity of such plugs and sockets. Low voltage dc electricity is a polarized power source. Lightbulbs are non-polarized appliances; it does not matter which side of the lightbulb is negative or positive. Not all devices are this way. Twelve Vdc electronics, motors, and fluorescent lighting are polarized. Hooking them up backwards can destroy them. Use a common polarity scheme for wiring the plugs and sockets and stick to it.

The major disadvantage to using 120 Vac outlets for low voltage is one of confusion. It is against most wiring codes to use the same types of outlets for two different forms of power (say 120 Vac and 12 Vdc) in the same building. Should a device be plugged into the wrong form of power, damage and even fire can result. If the 120 Vac outlets are used for low voltage dc, be sure they are plainly marked or color coded for this information.

There are other types of plugs and outlets which are polarized and suitable for low voltage dc use. The automotive cigar lighter plug is one example. The 20-ampere 120 Vac receptacle is another. There are other types of power plugs for 240 Vac service which have large contact areas. These plugs can be expensive. Keep their use to a minimum.

Whatever type of output receptacle and plug scheme is used, it will not be as efficient as directly soldered wiring. If all factors are considered, then soldering the devices to the line is the best method to use. It is cheaper, and more efficient.

INTERNALLY CROSS-WIRING THE BATTERY PACK

The interiors of large series-parallel wired battery packs may be cross-wired. This cross-wiring helps to distribute the loads evenly throughout the entire battery pack. This is desirable to minimize small differences between the separate cells which make up the battery.

Cross-wiring electrically puts parallel paths through the battery pack. These parallel paths help maintain a more even state of charge throughout the sub-batteries making up the pack.

This cross-wiring should be done with the same large wire used in the primary wiring of the battery pack. The same low resistance connectors should be used. Figure 7-1 shows the cross-wiring of a 12-volt 700 ampere-hour battery pack composed of four Trojan L-16 deep cycle batteries.

The use of cross wiring is not necessary. As can be seen from Fig. 7-1, both methods of wiring give working battery packs of the same capacity. The only difference is the parallel interconnecting

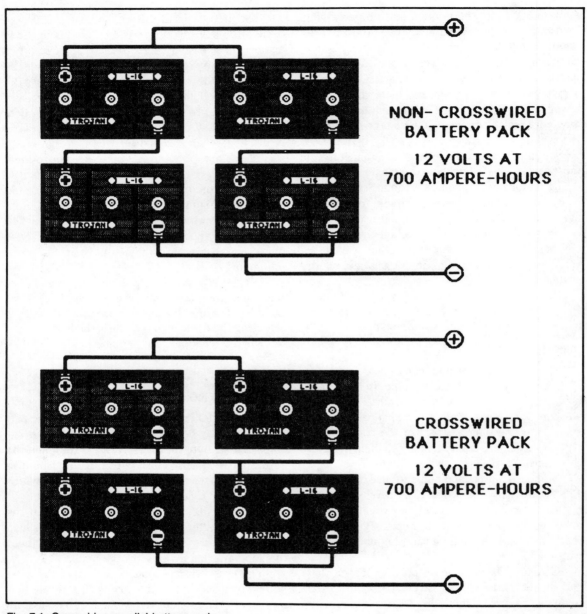

Fig. 7-1. Crosswiring parallel battery packs.

wire between the two halves of the pack. This wire gives each series string a parallel path to each other.

The cross-wiring technique is very useful in series strings of four or more sub-batteries (24- and 48-volt systems). In large series strings the failure of a single sub-battery can shut down the entire series string. The use of cross-wiring can prevent total shutdown of the entire string. It provides an alternative parallel path for the electricity. This parallel path does not pass through the weak sub-battery and therefore is not limited by it. The sub-batteries can be parallel cross-wired only at common voltage levels.

HOW TO USE WIRE FOR CURRENT MEASUREMENT

It is often necessary to measure the amount of current flowing in various circuits of a battery powered system. Information on how fast the battery is being charged or discharged is one example. We may wish to know how many amperes of current are being consumed by a particular device. By applying a simple technique, we can use the existing system wiring as shunts for current measurement. In this case, a shunt is a known quantity of resistance used to measure current in the form of a voltage drop.

The use of the existing wiring has two distinct advantages over other modes of measurement. The wiring is there anyway to transfer the power. The wiring does not need to be cut to make the measurement. When an in-line ammeter is used, the line must be broken in order to insert the meter. The ammeter has loss, as do the mechanical connections to it. Ammeters are expensive and introduce losses in the system's wiring that are not necessary.

Ohm's law tells us that the voltage drop across a piece of wire is directly proportional to the amount of current flowing through the wire ($E = IR$). We can determine the resistance of a particular length of wire by measuring its length and by looking up its resistance per thousand feet on the Copper Wire Table. The voltage drop across the length of wire can be measured by a voltmeter. The voltmeter should have a resolution to 0.1 millivolts (0.0001 volts). Digital Multimeters (DMM) are ideal for this task.

If we know the length of wire's resistance and can measure the voltage drop across it, we can calculate the amperage flowing through the wire by using ohm's Law.

$$E = IR$$
$$I = \frac{E}{R}$$

E = Voltage drop across the wire expressed in volts
R = Resistance of the piece of wire expressed in ohms (Ω).
I = Current flowing through the wire expressed in amperes

In actual practice the length of wire is chosen so that its resistance is a multiple of ten, that is, 1Ω, 0.1Ω, 0.01, 0.001Ω, or 0.0001Ω. If the resistance of the length wire is a multiple of ten, then the arithmetic of converting the voltage drop to an average figure is very easy and does not require a calculator. If the shunt has a resistance that is a multiple of ten, then the reading given by the voltmeter in millivolts may be used directly, with only the decimal point's position to be determined.

For example let's consider a current measuring shunt with a resistance of 0.0001Ω made from some 0-gauge copper wire. 0-gauge copper wire has a length of 10180 feet per Ω. By multiplying 10180 feet per Ω by $0.0001\ \Omega$, we can calculate the length of 0-gauge wire which has a resistance of 0.0001Ω. This is 1.018 feet or 1 foot and 7/32 of an inch. A voltage drop of 0.1 millivolts across this 1.018 feet of 0 gauge wire indicates a flow of 1 ampere. A drop of 1.0 millivolts across the 1.018 feet of 0-gauge wire indicates a current of 10 amperes, and so forth. In general, the length of wire needed to make any particular shunt is equal to the length of that gauge wire in feet per Ω times the resistance value in ohms of the desired shunt. In this case of 0 gauge wire:

$$L = (ft./\Omega)\ (\Omega)$$
$$L = (10180\ ft./\Omega)\ (0.0001\Omega)$$
$$L = 1.018\ feet$$

0 gauge wire is often used in connecting batteries to the bus. By measuring the voltage drop over 1.018 feet of this wire, we can measure very accurately the current flow to and from the battery.

Figure 7-2 illustrates using a digital multimeter (DMM) and a shunt of 1.018 feet of 0-gauge wire for current measurement. The DMM should be set to read millivolts. Each 0.1 millivolt of drop across the wire shunt indicates 1 ampere of current. The DMM will read positive voltages when current is flowing into the battery. It will indicate negative voltages when current is being used from the battery.

The shunt can be located in either the positive or negative wires. This is true of all shunts; it makes no difference whether it is in the positive or negative lead.

Note the use of a wooden beam with long pieces of threaded steel rod through it. This connection technique allows for multiple low loss connections to be made away from the battery. The battery only has two wires on its output terminals. This makes maintenance of the battery's connections easier.

Table 7-2 is a table of copper wire lengths used for making shunts of 0.1Ω, 0.001Ω, and 0.0001Ω. In general, use shunts of 0.1Ω for measuring cur-

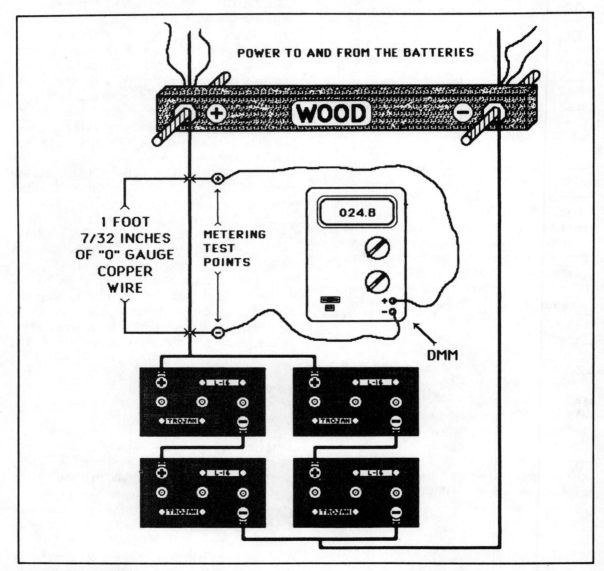

Fig. 7-2. Using shunts.

Table 7-2. Copper Wire Shunt Table.

WIRE GA.	MV. X 10 = MA. 0.1 OHMS		MV. = AMPS. 0.001 OHMS		MV. X 10 = AMPS. 0.0001 OHMS	
	FEET	METERS	FEET	METERS	FEET	METERS
0000	---	---	20.4	6.22	2.04	0.622
000	---	---	16.2	4.93	1.62	0.493
00	---	---	12.8	3.91	1.28	0.391
0	---	---	10.2	3.10	1.02	0.310
2	---	---	6.4	1.95	0.64	0.195
4	---	---	4.03	1.23	0.403	0.123
6	---	---	2.53	0.772	0.253	0.0772
8	---	---	1.59	0.485	0.159	0.0485
10	---	---	1.00	0.305	0.100	0.0305
12	---	---	0.63	0.192		
14	---	---	0.396	0.121		
16	24.9	7.59	0.249	0.076		
18	15.66	4.77	0.157	0.048		
20	9.85	3.00	0.0985	0.030		
22	6.20	1.89				
24	3.90	1.19				
26	2.45	0.747				
28	1.54	0.470				

WIRE TEMPERATURE 68° F. 20°C.

rents less than 2 amperes. Shunts having 0.001Ω can be used in current measurements up to 50 amperes. For over 50 amperes, use a shunt with a resistance of 0.0001Ω. The idea here is not to have the shunt introduce so much resistance into the circuit that it will affect the circuit's operation.

Any piece of wire anywhere in the circuit can be used as a shunt by following the aforementioned procedure. The shunt's resistance need not be an exact multiple of ten, if a calculator is used to convert the voltage measurement into the ampereage figure. A single, very accurate voltmeter can be used for all metering chores within the system.

The use of copper wire for shunts depends on the accuracy of the Copper Wire Table. Wire of a certain gauge size must have the resistance specified on the table. In most cases this is true. The manufacture of copper wire is held to very strict quality control. The gauge wire used will have the resistance per foot as specified on the table.

One problem with using copper wire for shunts is that it changes in resistance due to temperature. If the wire is cold it will have less resistance than if it is hot. The Copper Wire Table is calibrated for a wire temperature of 68° F. (20° C.). If the wire is at 32° F. (0° C.), then its resistance will be about 10 percent lower than the 68° F. figure listed on the table. If the wire is at a temperature of 122° F. (50° C.), its resistance will be 10 percent higher than the figure on the table. If copper wire is used for very accurate current measurement, its temperature should be taken into account.

THE BATTERY'S LOCATION IN THE SYSTEM

Where the batteries are physically located in the system is of importance. Safety is the prime factor. The batteries should also be located to minimize loss from two other factors—length of wiring and temperature.

The battery should be located in the center of the bus rather than at either end. This central location reduces the loss to branch circuits at the end of the bus. If possible the building should have a *battery room* which is centrally located.

Another factor in the battery's location is temperature. The batteries should be held at a constant room temperature if at all possible. Basements are good locations because of their relatively constant room temperature. Banishing the batteries to the back porch or to an unheated garage or shed leads to inefficiency from extra wiring and low temperatures.

Batteries must also be located where they do not present a hazard to humans, especially small ones. As such, safety is a prime factor in the location of the batteries in the system and in the building.

SAFETY REQUIREMENTS

Batteries are filled with dangerous chemicals. The acid or basic electrolyte solutions can burn human flesh and even blind you if they get in your eyes. Be careful when handling the electrolyte. Always keep several gallons of water available to dilute any electrolyte spills. Baking soda dissolved in water is useful in neutralizing the sulphuric acid electrolyte in lead-acid cells. Vinegar will neutralize the potassium hydroxide electrolyte in vented nickel-cadmium cells. Keep these neutralizing agents away from the tops of the batteries. Cleaning the batteries' tops with neutralizing solutions is a shortcut to contamination and ruined batteries.

Keep batteries away from children.

Under no circumstances should young children have access to the battery room or area. Not only is there a danger of chemical burns, but the electrical energy stored within the battery is dangerous. If the batteries are short circuited by, say, a wrench, this wrench will become hot enough to burn in less than a second. Children may see a parent maintaining a battery and decide that this is an activity for play. Inform children of the dangers of the batteries. If young children are present, make sure the battery area stays locked.

Ventilation is necessary in the battery area. All vented batteries emit explosive gases when they are charged. Be sure the area is well ventilated. No smoking or open flames should be allowed in a non-ventilated battery area. If motorized ventilation is used in the battery room, it should have sparkless motors.

Don't make live connections directly to the battery. The spark which occurs when a live connection is made to the battery places a source of ignition in the area most likely to contain explosive gases.

Battery rooms which are well ventilated, especially during charging, are not likely to be hazardous. It's better to be safe than sorry.

BATTERY INSTRUMENTS

In order to use batteries effectively, we must be able to measure and evaluate their performance. A few simple measuring tools are necessary if the batteries are to be run efficiently. The minimum instrumentation necessary is a system voltmeter and an ammeter. The ammeter measures current flow to and from the battery. both these tasks can be performed by a single voltmeter if wire shunts are used.

The Voltmeter

The voltmeter used to measure system voltage must have high resolution over the voltage range covered by the battery. Analog voltmeters are usually referenced to zero, which makes a large portion of the scale unused. For example, in a 12-volt battery system, the voltage varies between 11 and 16 volts. The system voltage is never below 11 volts nor greater than 16 volts. If a meter with a scale of 0 to 20 volts is used, the area of the scale from 0 to 11 volts is not used. Neither is the portion from 16 to 20 volts. In effect, the meter's resolution is reduced by 75 percent. Any analog voltmeter that is used to measure system voltage should be of the expanded scale type.

If the same voltmeter used to measure system voltage is to be used with shunts for current measurements, then it must also have resolution to 0.1 millivolts on the lowest voltage scale. Most analog meters simply do not have the resolution and sensitivity to perform both jobs. This job is best done by the digital multimeter.

The Digital Multimeter (DMM)

The DMM is a very basic piece of equipment necessary to any battery or alternative energy system. It will perform measurement of voltage, current and resistance. It is invaluable in troubleshooting battery systems.

The DMM is a digital device; it reads in numbers, not via a pointer and scale like analog meters. The DMM's accuracy is very high. It is easy to use and relatively rugged compared to analog meters.

The DMM no longer costs more than the analog types. If accuracy is considered, it is much cheaper. Digital multimeters cost between $50 and $150 depending on accuracy and quality.

The Micronta DMM, sold by Radio Shack (RS #22-191), is adequate and available for $59.95. The Fluke DMM, model 77, has better resolution and accuracy. It is very rugged and costs about $130. Only consider DMMs which have resolution to 0.1 millivolts or less on the lowest voltage scale. This ensures adequate resolution for shunts with as little resistance as 0.0001 Ω.

The DMM can be used to measure system voltage, individual cell voltages, amperage by using shunts, and ac output from inverters. Without some form of accurate, reliable measurement, the alternative energy user is blind. This type of accurate information is essential to proper battery use.

Expanded Scale Analog Voltmeter

The resolution of an analog meter can be increased by expanding its scale. A small electronic circuit can be built to make the meter start reading at a point other than zero. The full scale of the meter can be used to indicate the desired voltage range. This effectively increases the resolution of the analog meter. If a sensitive meter is chosen (100μa. or less), the meter can be left on line all the time. It will draw very little power and constantly read out battery voltage.

Figure 7-3 gives a schematic for an expanded scale voltmeter. This circuit is designed to work on a 12 Vdc system. The meter's scale will start reading at 11 volts. It will be full scale at 16 volts. The meter used is a dc ammeter with a full scale reading of 100 microamperes (μa). This dc ammeter is converted into a 11- to 16-volt voltmeter by the electronic circuit. Leave the leads on the zener diode long for heat dissipation. An expanded scale meter of this type gives about the same resolution as a 3 1/2 digit DMM. The device should cost less

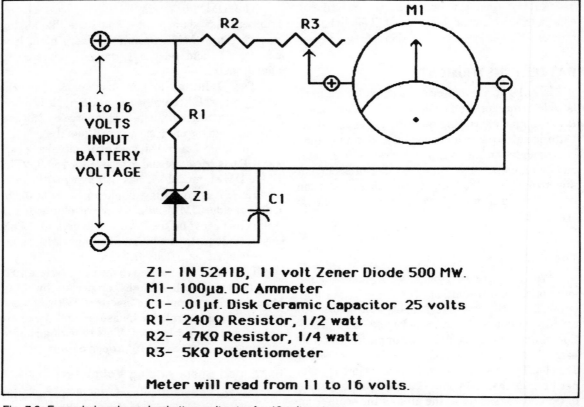

Fig. 7-3. Expanded scale analog battery voltmeter for 12-volt systems.

than $10 to make if a surplus meter is used.

Similar metering tricks can be performed on 24- and 48-volt battery systems by adapting the circuit to the higher voltages. An accurate battery voltmeter is essential to running any battery system. This is especially true of systems using lead-acid batteries for storage.

The Hydrometer

The hydrometer is of limited benefit to users of lead-acid batteries. By using the hydrometer to measure the specific gravity of the electrolyte, the state of charge of the battery can be determined.

The only type of hydrometer to consider has built-in temperature compensation. This type is commonly available at automotive stores for under $10.

If a hydrometer is in use, be sure it is clean. Consult the chapter on lead-acid batteries for the results of cell contamination. Considering that the battery's state of charge can be determined by its voltage, usage of the hydrometer is discouraged. There is simply too much danger of contamination.

FILTRATION--KEEP THE NOISE OUT

A commonly encountered problem in battery systems is noise. Many devices such as motors, computers, and power control circuits can introduce noise to the main bus. This noise can be detected and amplified by radios, televisions, and stereos.

It is difficult to totally eliminate noise from battery systems. The reason for this is that all the devices share a common ground through the negative pole of the battery. Each noise-making device can have its power input leads bypassed with a 0.1 microfarad (µf) disk ceramic capacitor. If this is not effective, consider putting a choke in series with the power leads of the device. The choke can be put in either the positive or the negative lead. Try both and use what works best. Be sure that the choke has adequate current handling capabilities for the job.

In many cases the noise can be eliminated on the bus by the addition of a large amount of capacitance. Connect from 50,000 µf. to 500,000 µf. across the main power leads from the battery to the bus. Use electrolytic capacitors, and be sure the polarity and voltage rating on them is observed. This will reduce the ripple on line and filter out most of the noise. At the same place it is also effective to put a 0.1 µf. disk ceramic capacitor to filter very rapid noise pulses.

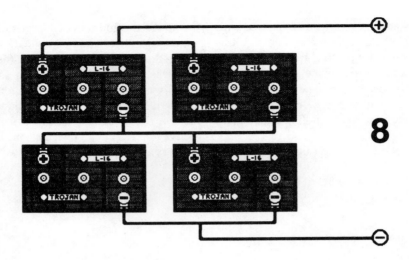

Inverters

An inverter is a device which converts direct current (dc) electricity from the battery into alternating current (ac) electricity. Inverters come in many sizes and types, but they all convert dc power to ac power. Usually the inverter also steps up the voltage of the dc input power to a higher output ac voltage. The inverter makes it possible to convert the battery's dc energy into regular 120 Vac, 60 cycles per second housepower.

The use of inverters has revolutionized the field of alternative energy. For the first time, battery users can have standard 120 Vac housepower constantly available without running a motorized generator. The inverter makes it really practical and inexpensive to be one's own power company. An inverter can provide quiet and efficient standard 120 Vac power in areas where commercial housepower is either not available or too expensive.

The inverter also makes it possible to use energy stored in batteries as a backup for commercial power. Batteries through the inverter can power standard 120 Vac equipment in the event of commercial power outages. Airports, hospitals, communication centers, and computer users are examples of services that must have uninterruptible ac power.

Direct current (dc) electrical energy is the only type available from the battery. The voltage and quantity of this dc energy depends of the configuration and size of the cells that make up the battery. Dc battery energy will not directly run transformers or ac type motors. The energy directly from the batteries will not power most standard household appliances. The dc energy from the batteries is inherently low voltage, yielding high losses in the wiring. All these deficiencies of battery power can be eliminated by using an inverter in the system.

HOW THE INVERTER WORKS

The inverter takes the dc energy from the battery and "chops" it into pulses of electrical energy. These pulses have about the same voltage as the battery. The pulses are then fed into a transformer where they are increased in voltage. The output of the inverter's transformer is an alternating current of higher voltage. Hopefully, the inverter has an output power which simulates the standard 120 Vac, 60 cycles per second, power available from the power company.

Switching

The chopping or switching of the battery's energy is accomplished by transistors or silicon controlled rectifiers (SCRs). The transistor type of switching is more efficient; the SCR is cheaper. The transistors or SCRs are controlled by the inverter's logic section. The logic section provides a switching signal of 60 cycles per second to drive the transistors or SCRs. The logic section also regulates the output voltage of the inverter to the proper level.

Once the battery's energy is chopped, it is essentially an alternating current (ac) form of electrical power. Alternating current (ac) electricity has the property of a constantly changing direction of the electronic flow. This constantly changing current produces a constantly changing magnetic field. This property of ac power allows it to be changed in voltage by using a transformer.

Transformer

The transformer uses the constantly changing magnetic fields generated by ac power to transfer energy. Transformers are only operable on ac electrical power. The dc energy from the battery will not directly operate a transformer. The dc must first be chopped into pulses before it can drive the transformer.

The transformer is capable of raising the voltage of its input power. In a 12-volt system, the inverter's transformer is fed ac pulses of about 12 volts and the transformer puts out pulses of about 120 volts. The 12-volt pulses are applied to the input (primary) of the inverter's transformer. These pulses induce a changing magnetic field within the transformer. This magnetic field, in turn, induces a current in the output (secondary) windings of the transformer. The ratio of the number of turns between the primary and the secondary determines the amount of voltage transformation. If the primary has 100 turns and the secondary has 1000 turns, then the ratio is ten. The output voltage of such a transformer will be 10 times the input voltage. If 12 volts is fed to the primary, then 120 volts will be available at the secondary.

Now, this may sound suspiciously like the elusive "free lunch". Twelve volts in one end and 120 volts out of the other. There is, however, no free lunch to be found here. The power output of a transformer is less than the input power. In the inverter's case, the transformer changes voltage at the expense of current. For example, let's say the transformer is transferring 120 watts of power. It will take 10 amperes of current at 12 volts to drive the input of the transformer. The transformer's output will be 120 volts but at less than 1 ampere (P = IE or 120 watts = 120 volts times 1 ampere). Actually the transformer is slightly less than 100 percent efficient. Some power is lost in the transformation process. Energy is not created within the transformer; it is merely transformed. Voltage is increased at the expense of current.

The transformer has been around for a long time. It was essentially discovered by Michael Faraday in 1831. It is only since the development of the silicon junction, i.e. transistors and SCRs, that the transformer has been successfully applied in inverter circuits. Silicon technology has made possible the switching circuits necessary to chop the dc energy efficiently. Efficient and reliable power inverters are a relatively new development.

DIFFERENT TYPES OF INVERTERS

Inverters are manufactured in many types. They differ in order to be able to do a wide variety of power chores. If inverter application is to be successful, then the proper type must be used. They are made in a variety of power levels, since they are most efficient if properly sized to the job at hand.

One major difference between inverters is the type of output power they produce. The basic question is how closely the inverter's power resembles the standard commercial power. Commercial power has a sinusoidal waveform (sine wave) with a frequency of 60 cycles per second. In many types of equipment, a sinusoidal (or a close approximation to it) input power is necessary. Other types of devices can use power which is not sinusoidal.

The output power of inverters can be classified into three basic types: square wave, modified sine wave, and pure sine wave. Figure 8-1 shows examples of the different types of output waveforms produced by each type of inverter. Each type has different operating characteristics and applications.

Square Wave Inverters

The square wave inverter's output is least similar to commercial power. As such, the range of devices operable from it is the smallest. This type of inverter is designed to provide power for the cheapest possible price.

These inverters use the minimum number of parts necessary to function. The logic sections are not very sophisticated. Often the output frequency of this type of inverter will vary with the battery's voltage. As the battery runs down, the inverter's output frequency slows down. This variation in frequency can make the power unsuitable to many types of appliances. Motors, television sets, record turntables, and tape recorders all require their input power to be very close to 60 cycles per second.

Square wave inverters also provide very poor voltage regulation. The output voltage may vary as much as 30 volts. This voltage variation is a function of the changing input voltage to the inverter as the battery discharges. Variations of over 10 volts in the inverter's output can lead to trouble operating many types of equipment.

The square wave inverter is only efficient if used at or near its rated output power. Efficiency at full output is around 80 percent. At lesser output levels the efficiency drops radically. For example, let's consider a 1000-watt square wave inverter. At an output of 1000 watts, this inverter is about 80 percent efficient. At an output of 500 watts, the efficiency drops to less than 50 percent. At an output of 200 watts, the efficiency is less than 30 percent. This low efficiency when lightly loaded makes

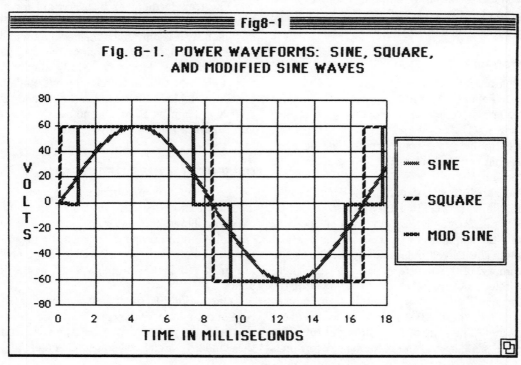

the square wave inverter a poor choice for household alternative energy service.

The logic sections of square wave inverters make no provision for the amount of energy being consumed. The inverter is always consuming large quantities of power even when no power is being used from it. These units cannot be left on, waiting to be used. Efficiency dictates shutting them off when not in use.

Square wave inverters are actually only suitable for powering resistive loads like soldering irons, heaters and incandescent light bulbs. They will drive most brush type motors. They are not suitable for electronics, video equipment, computers, or printers. Fluorescent lighting used with them will work, but the light emits an annoying buzzing sound. Fluorescent light ballasts wear out very quickly when powered by square wave inverters.

Square wave inverters are available in output wattages from 30 watts to 1000 watts. They are usually set up for 12-volt input. Prices vary from about $50 to $800. Considering that the modified sine wave type is only slightly more expensive, the square wave inverter is not cost effective. It is suitable only for small simple jobs such as powering an electric razor in an automobile.

Modified Sine Wave Inverters

The modified sine wave inverter is the type most commonly used in alternative energy systems. These units have become very efficient and ultra reliable. The modified sine wave inverter will run almost all 120 Vac powered appliances and electronics. These inverters are made in sizes large enough to power an entire household. In many cases, the inverter system is automated to the extent that it requires no human management.

The output of these inverters is a modified sine wave. The output waveform is shaped midway between the square and the sine wave. This waveform is close enough to a sine wave to fool most appliances. The modified nature of the wave makes it possible to use very efficient switching techniques within the inverter. The modified sine wave inverter represents a livable compromise between utility and efficiency.

This type of inverter's output is both frequency and voltage regulated. The output power remains very close to 120 Vac 60-cycle power over the entire input voltage range. The inverter's output is stable over the entire battery's capacity.

The modified sine wave inverter is usually equipped to sense the amount of power being used. It will adjust its consumption to meet the demand for power. If there is no load present, it will reduce input power to save energy. This type of inverter is capable of responding to light loads with less input power. These inverters are overall much more efficient than any other type. It is possible to leave them on all the time without wasting large amounts of power.

As an example, let us consider a 1,2000-watt inverter which is very popular with alternative energy users—the Heart HF 12-1200X. This inverter represents the state of the art in inverter power and is manufactured by Heart Interface, Federal Way, WA. The frequency stability is one percent at 60 cycles. The output voltage is regulated within two percent. The amount of energy consumed by the inverter in a no-load condition is less than 1 watt. Continuous output power is 1,2000 watts, with a surge capability of 3,5000 watts. At an output of 200 watts, the inverter's efficiency is around 93 percent. The inverter's input power is a 12-volt battery. These specifications apply to a voltage input range between 10.8 volts and 15 volts. This inverter uses super-efficient field effect transistors (FET) as switching devices.

The logic sections of this inverter are very sophisticated. The inverter senses the amount of load present and adjusts itself to meet the demand. The output power regulation is actually tighter than that of commercial power. The logic even contains a circuit that shuts the inverter off when the battery is close to fully discharged. This saves the battery from over-discharge and premature death.

The efficiency of an inverter is of critical importance. Most inverters spend much of their time being used at much less than full output. Efficiency must be as high as possible over the entire operating power range of the inverter. Figure 8-2 shows the efficiency versus output power for the

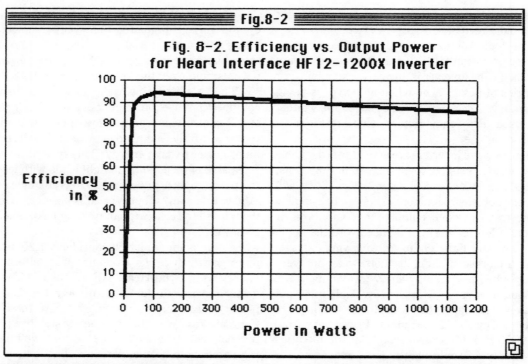

Fig. 8-2. Efficiency vs. Output Power for Heart Interface HF12-1200X Inverter

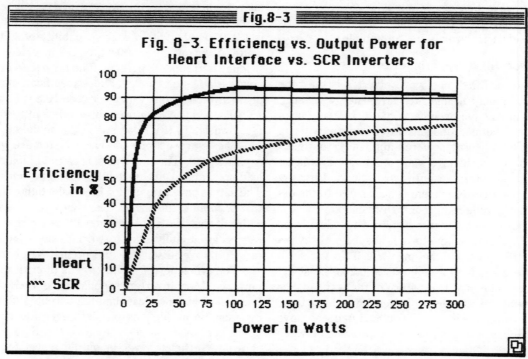

Fig. 8-3. Efficiency vs. Output Power for Heart Interface vs. SCR Inverters

Heart 1,200 watt inverter. Figure 8-3 shows efficiency vs. output power at low levels for two types of inverters. One uses SCRs for switching devices; the other uses transistors.

The modified sine wave type of inverter is capable of powering almost any device that will run on standard commercial power. It will run computers, printers, satellite television systems, fine audio gear including turntables, motors, power tools, washing machines, and refrigerators. The list is literally endless. Some types will not run very delicate equipment such as biomedical electronics.

The modified sine wave inverter is making it possible to run a normal electrical household from batteries. Figure 8-4 illustrates the usage of a 1.2 kW. inverter in a battery system. This illustration shows a number of power inputs: alternative energy, motorized dc, and motorized ac. Any one of these inputs will suffice to establish a primary system.

The modified sine wave inverter is available in many sizes, from 300 watts to 12,000 watts. Most are capable of three times their rating for short surges. The larger inverters are capable of 240-volt output. Export models are available with 240 Vac, 50-cycle per second output. The prices vary widely from manufacturer to manufacturer. A 1,200-watt inverter costs between $1,000 and

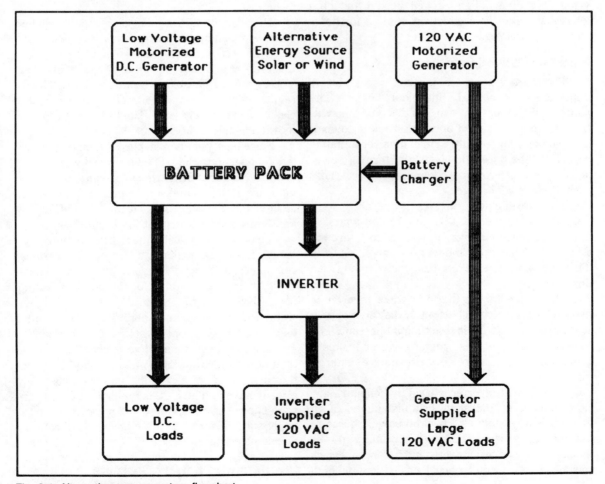

Fig. 8-4. Alternative energy system flowchart.

$1,600. The 5,000 watt models cost around $2,400 to $3,300. These inverters cost about $1 per watt. The larger inverters use 24- to 140-volt storage battery packs as an input power source.

Those considering inverter power should research the different brands that are available to them. This type of inverter has undergone considerable technical refinement within the last few years. Cost and efficiency vary greatly from brand to brand. While the inverter is a poor component to economize on, spending more money does not mean getting the most efficient inverter.

The modified sine wave inverter is very reliable. They put out power 24 hours a day for years at a time. Their rugged design and high longevity make them excellent choices for boats and other vehicles.

Pure Sine Wave Inverters

The pure sine wave inverter most closely duplicates commercial sinusoidal power. The nature of the output wave makes this type inherently inefficient. It is used only as a backup to commercial power where ultra clean inverter output is essential. The pure sine wave inverter will power any device which requires commercial 60 cycles per second sinusoidal power.

The pure sine wave inverter is available in output powers between 1,000 and 15,000 watts. A 1,000-watt pure sine wave inverter costs around $3,000. A 5,000-watt model costs about $5,200. This type of inverter is more expensive to buy than any other type.

The efficiency of the sine wave inverter is around 60 percent at full output. If they are not fully loaded, the efficiency is less than 30 percent. The lower efficiency of this type of inverter makes it unsuitable for usage in alternative energy housepower.

This type of inverter is used by hospitals, airports, and large industrial establishments. In all cases it is used only as a backup power source to commercial power. It is not a primary power source. It is used infrequently and for short periods of time. This type of service does not demand high efficiency.

Uninterruptible Power Supplies (UPS)

The uninterruptible power supply (UPS) is a specialized type of pure sine wave inverter developed for powering computers and computer equipment. The UPS is a short term backup for commercial power. It can react to power outages fast enough to keep data from being lost in computer systems.

The UPS will react to a power outage in less than 10 milliseconds. The other types of inverters take about 1 to 3 seconds to come on line after a power outage. In most applications, a lapse of 3 seconds on power is not critical. In a computer, a lapse as short as 100 milliseconds can cause data to be lost.

High quality UPSs are also equipped with line filters to clean up the power from the power company. They filter the commercial input power when it is present and they are not in use as an inverter. The UPS has a sinusoidal output that is very closely regulated. Their power output is more reliably clean and better regulated than the power supplied by the electric utility.

Since most commercial power interruptions are less than an hour, the UPS is equipped with a small capacity battery. Often this is a sealed battery to interface with the office environment. Their use is so intermittent and short that their low efficiency and limited capacity is not important.

They are primarily used by computers in banks, hospitals, airports, and other professional computer applications. Recently, the UPS has been available in sizes suitable for the home computer user. UPS sizes vary from 250 watts to 15,000 watts. Prices are about $4 per watt for the smaller ones to about $1.20 per watt for larger models.

Most users of alternative energy have some type of inverter running 24 hours a day. As such, they don't need uninterruptible power because they already have it. The UPS is specialized for use only as super clean backup to commercial power. It is not suitable for alternative energy applications, due to cost and inefficiency.

Inverters with Battery Chargers

The modified sine wave inverters and pure sine

wave inverters are available with a battery charger built in. This charger makes it possible to charge the battery pack from 120 Vac commercial power. This feature is primarily intended for standby service—as a backup for commercial power. However, the battery charger can be used to establish a primary energy system with the addition of a motorized ac powerplant. This is probably the simplest and most common form of alternative energy power used in household service.

For example, consider a backwoods homestead which already has a motorized ac powerplant. There is ac power available only when the plant is running. If an inverter/battery charger and batteries are added to the system, the ac power becomes continuous without having the motor running. The system becomes much more efficient as the motor/ac generator spends most of its time off. Such a system costs less than 20 percent to operate as compared with continuous generator operation. It is common to run the generator only for about a day a week to charge the batteries.

The battery charger within the inverter senses the presence of the generator's power and begins to charge the battery pack. The inverter also automatically transfers all ac loads to the generator. When the generator is turned off, the inverter automatically transfers the loads back to the inverter and begins making ac power again. Changeover time is in the neighborhood of 2 seconds.

The battery charger within the inverter is designed for standby or float service. In float service the battery pack is continually being charged, 24 hours a day, every day by commercial power. The control logic on the battery charger is set for float service. The voltage limit is usually far too low for efficient deep cycle use. The current output of the battery charger is often very small. In float service, we have days to complete the charge process. There is no need for speed. This is not the case when the power source is a motorized generator. We want to charge the battery pack as rapidly as possible to minimize the running time of the generator.

Inverter/battery chargers can be used in deep cycle alternative energy systems if they meet two requirements. The voltage limit of the battery charger must be adjustable to allow for more rapid charging of the battery. The battery charger must be capable of delivering enough current to charge the battery pack at the C/20 rate. If the battery charger is capable of meeting these requirements, then it will work as the only charge source for the battery pack. If there are additional charge sources, then these requirements are not critical but merely desirable.

The inverter/charger, the motorized generator, and the battery pack represent the minimum equipment necessary to establish a stand alone alternative energy system. This is the cheapest way to have continuous ac power available without buying it from the power company. The cost of this setup is many times cheaper than either solar or wind power of the same capacity. If one considers fuel cost and engine maintenance, this system is still two to four times less expensive than wind or solar power.

PROPER INVERTER SIZING

The inverter must be properly sized to efficiently meet energy demands. If the inverter is undersized, then there will not be enough power. In most inverters, demanding more than their limit will shut them off. If you overload them, all the lights go out. If the inverter is oversized it will be less efficient and more costly to buy and to run.

In order to properly size an inverter one must first know the amount of energy that is required. This is a complex task since energy consumption is not constant. Chapter 9 covers the details of load estimation and power consumption estimation. In inverter sizing the most important factor is peak power consumption. The peak power demand should not exceed the rated peak output of the inverter.

Estimation of peak power demand is difficult when it is possible for many devices to consume power at the same time. The problem is further complicated by any electric motors in the system. Some types of electric motors require three times as much power to start them as is required to run them. If two or more motors are started at the same time the peak power demand is much higher than the average demand.

The inverter should have at least enough power output to be able to start the largest motor in the system. In most systems this represents the minimum peak power consumption. If a small inverter is to be used to power several large motors, such as a well pump and a refrigerator, care must be taken that they do not start at the same time. It is possible to put one or more of the motors on manual on-off control by eliminating the automatic control. If two or more motors are left in the auto-start mode, the inverter must be sized to simultaneously start them all. If this is not done, the inverter may shut itself off due to overload. Most inverters are capable of starting surges of three times their rated output.

A one kilowatt inverter can supply power to an energy conscious household if there are no large motors in use. Two and a half kilowatts is the minimum size that will support large appliances such as washing machines and refrigerators. Deep well pumps and other large motors are usually powered by 5 kilowatt or larger inverters.

In most systems peak power demand exceeds average power usage by four to ten times. Proper energy management can reduce peak demand. The basic idea is this: don't turn everything on at the same time. If energy management is used, the inverter can be sized closer to the average power demand. This will increase the system's efficiency and reduce hardware costs. For those not wishing to practice energy management, the system is much more expensive. The inverter must be grossly oversized in relation to average power consumption. Such a system will be less efficient to run and more expensive to buy.

TYPES OF BATTERIES SUITABLE FOR INVERTER USE

The type of battery which is suitable to power an inverter depends on two factors: the output power of the inverter and the type of service in which it is being used. Uninterruptible power supplies are usually powered by lead-acid gel cells or by sealed sintered plate nickel-cadmium cells. Larger inverters used in alternative energy systems use deep cycle high antimony lead-acid cells or vented pocket plate nickel-cadmium cells.

Automotive batteries are not suitable power sources for large inverters. The inverter is capable of demanding hundreds of amperes from the battery during surge conditions. The battery must be able to deliver energy fast enough to meet the demands of the inverter. The only types of batteries that are capable of doing this for extended periods of time are the true deep cycle lead-acid battery and the vented pocket plate ni-cad battery.

The output voltage of the inverter is regulated. As such, the voltage fluctuation of the lead-acid cell during cycling is less important than if the energy was being used directly from the battery. Lead-acid cells are the most cost effective type for powering inverters.

The battery pack must have enough capacity to run the inverter at full output without too much battery voltage depression. Ideally the battery should have the ability to power the inverter to full output for at least ten hours. For example, consider an inverter with an output of 1 kilowatt. The battery must be capable of 10 kilowatt-hours of service (1 kilowatt × 10 hours = 10 kilowatt-hours). Assume that the inverter is being powered by a 12-volt battery pack. Ten kilowatt-hours divided by 12 volts is 833 ampere-hours. This is the minimum capacity battery that can efficiently power the inverter without excessive voltage depression. There are, however, other considerations in sizing the battery's capacity. Chapter 9 discusses the sizing of battery packs to meet specific energy needs.

INVERTER TO BATTERY INTERCONNECT

The wiring and connections between the battery and the inverter are critical areas. The inverter must have a very low resistance path to the battery's energy. All wiring must be of large size, and all connections must be impeccable. Currents over 200 amperes will be flowing through this circuit. Excessive loss in the wire or connections can significantly reduce the inverters ability to respond to surges, and will lower the system's efficiency.

The wiring lengths between the inverter and the battery should be kept to an absolute minimum. In most installations, wire runs of over 5 feet in length between the battery and the inverter are considered long. The wire should have a minimum size of "0" gauge copper. If the inverter is 5 kilowatts or larger or if the wire runs are long, use "0000" gauge copper wire. Low loss connectors should be used on the wire ends.

In most cases switching is not used to the inverter's input. The inverter is connected directly to the battery or to the battery bus. Switches which have low enough losses to pass 200 amperes or more are rare and expensive items.

All connections between the batteries and the inverter should be kept clean and bright. This is very important. Corrosion is inevitable and periodic maintenance is the only solution.

INVERTER LOCATION

The losses involved in transferring low voltage electricity make it necessary to situate the inverter as close to the battery as possible. This is hardly an ideal environment for the inverter. The area around the batteries is likely to contain corrosive gases and liquids. The inverter should be isolated from the battery in every way possible except distance.

Ventilation is essential to the inverter. These devices produce quite a bit of heat when they are working at near their rated output. Heat is the number one killer of semiconductors. If the inverter is to last, it must be kept as cool as possible. If the ambient temperature is high, put a fan on the inverter. Do all that is possible to ensure the inverter's temperature remains as low as possible.

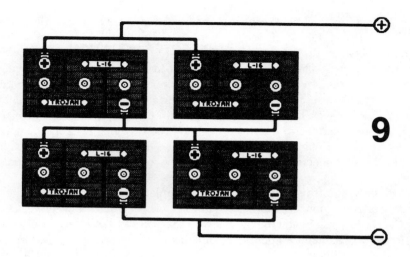

Energy Management

Energy management means getting your money's worth. Energy management can make the difference between an expensive system that does not satisfy and an efficient low cost system which can meet the need. The process of energy management is not difficult, but it does require intense attention to detail.

Energy management is the estimation of energy consumption, proper sizing of equipment to meet this estimate, and proper operation of the equipment. Energy management also deals with the selection of appropriate and efficient appliances. The time to start practicing energy management is before the system is purchased. Estimation of energy consumption is the first and most difficult step. It is also the most important step. Energy needs dictate the size and type of battery. Need also determines the amount of charging energy required.

HOW TO ESTIMATE ENERGY CONSUMPTION

An accurate energy use estimation is essential to all decisions made in systems with batteries. This is more than merely adding up the total power consumption of all appliances. In alternative energy systems the power estimation is commonly expressed in watt-hours per day (W-hrs./d).

A major factor in energy estimation is psychological. It is a common misconception that alternative energy systems provide free power. This is very far from the truth. Any alternative energy system capable of powering the average American household is going to be expensive. Parts wear out; there are operating expenses. At this point in time, alternative energy systems supply energy at many times the cost of commercial power. This will not always be true, but it certainly is now. Those considering alternative energy need to be discriminating in which appliances are electrically powered. Appliances should be selected with efficiency as a prime criteria.

Most appliances used in the average household are powered by electricity. Large heating devices like electric stoves, water heaters, clothes dryers, and space heaters are unsuitable for battery power.

They simply require too much power. This is not to say that they cannot be battery powered: it is just not cost-effective. In the alternative energy household, all heating chores can be handled at a fraction of the cost by burning propane, natural gas, or wood.

If heating devices are excluded, the next largest consumers are motors. Motors are used in refrigerators, washing machines, well pumps, vacuum cleaners, and a wide variety of small kitchen appliances. Motors can be run in alternative energy systems with complete success. It is a matter of proper sizing of the batteries and inverter. They are, however, large consumers of electricity so their usage should be held to a minimum. Care must be taken to select motors with the lowest possible startup surge.

Devices such as lighting and electronics are relatively small consumers of electricity. These types of appliances are prime candidates for battery power. In some cases an inverter is not necessary in the system. There are many types of electronics and lighting available for 12-Vdc power.

It is necessary to separate those jobs which must have electricity from those that do not necessarily have to be powered electrically. In an alternative energy system, it is cost effective to keep electrical usage to a minimum. In the words of the philosophers: *"It is a wise man who knows the difference between what he needs and what he wants."* While this may sound trite, it is certainly true in alternative energy systems.

Appliance Power Consumption

Most appliances have the amount of power they consume listed on them. This figure represents the peak power consumption of the device. If the device is not marked with its power rating, it can easily be measured. Measure the current being drawn by the appliance and multiply it by the input voltage. Power is the product of current times voltage ($P = IE$).

The total power consumed by a device is the product of its wattage times the amount of time it is operating. This amount is expressed in watt-hours (Wh), or kilowatt-hours (kWh). For example, if a 100-watt lightbulb is turned on for a period of 4 hours, it will consume 400 watt-hours of electrical energy. A 1000 watt refrigerator that spends 5 hours a day running will consume 5000 watt-hours of energy per day.

The difficulty in estimating the energy consumption of an appliance is not how much it draws, but how long it is on. Electrical usage varies greatly from season to season. The accurate energy estimate is made from worst case scenarios on the theory that slightly too much energy is better than not enough. Some devices such as lighting are predictable; other devices such as refrigerators are more difficult to estimate.

Lighting usage should be estimated at an average of about 5 hours per day. Each member of the household should be allowed a light. For example, consider a household of four persons. Each person is burning a 60-watt lightbulb for 5 hours per day. The power consumed by a single 60-watt bulb for 5 hours is 300 watt-hours. Since there are four bulbs burning at the same time the total consumption is 1,200 watt-hours per day for lighting.

Each appliance must have its wattage and duration of operation accurately measured. This is necessary in order to obtain an accurate estimation of power usage. Be sure to include all appliances in the estimate, even ones which are used infrequently.

In small systems using inverters with 12-volt inputs, it is desirable to separate the estimates of the different types of loads. There should be a listing for the 12-volt dc operated devices, and another for the devices powered with 120 Vac through the inverter. This is done to ensure that the inverter is sized with enough capacity to supply the ac loads and to include inverter inefficiency in the estimate.

Peak Power Consumption

The peak power demand is really only important to the users of inverters. The dc power from any quality battery storage system is capable of tremendous surges in relation to its capacity. This is not true of the inverter. The inverter is usually capable of surges between two and three times its rated output.

Peak power consumption is estimated by adding the wattage requirements of all devices that are to be operated at the same time. Peak power consumption is rated in watts. In this estimate the amount of time the appliances are on is not a factor. The peak power estimate is merely the sum of the wattages of all loads on line at the same time. The time of peak consumption usually occurs at night due to lighting.

Peak power consumption occurs when the largest number of devices are consuming power at the same time. The inverter is only capable of delivering so much power. If its rating is exceeded it will shut itself off. If all the lighting in the system is ac powered, then inverter shutdown will turn off all the lights. The estimation of peak power demand is essential to proper inverter sizing.

The starting surges involved in starting electric motors should be considered. Some types of electric motors draw three times their rated wattage when starting. If several motors are on automatic control it is possible for the starting surges to be coincident. A motor with a high starting surge may exceed the rated capacity of an already heavily loaded inverter. It is best to have only one motor in the system on auto-start. Usually this is the refrigerator. All other motors in the system should be turned on manually to reduce peak power consumption.

Peak power consumption may be very high when the starting surges of motors are considered. These surges last only for fractions of a second. As such, they have very little effect on the average power consumption of the system. They must, however, be allowed for to prevent overloading the inverter.

Average Power Consumption

In alternative energy systems, average power consumption is usually estimated on a daily basis. It is the amount of power consumed from the battery by all devices during a days time. This amount is usually larger in the winter than in the summer. If there are any doubts as to the amount of time a device spends on, estimate high.

The estimate of average power consumption should contain a margin for system expansion. The battery pack can easily be enlarged with the addition of more batteries. This is not the case with the inverter which must be replaced with one of higher output.

Average power consumption per day is the estimate used to size the capacity of the battery pack. If the system is to store enough energy for 10 days without power input, then the capacity of the battery must be 10 times the daily average power consumption.

Energy Estimate for a Small System

This section details the complete estimation of a hypothetical small alternative energy system. This example is for a household of two people, living in the complete absence of commercial power. It represents the use of electricity only where it is absolutely necessary. This example may be considered as the minimum system that is really effective and livable.

In this system, much of the energy is consumed as low voltage dc power used directly from a 12-volt storage battery. The inverter is used only for equipment that absolutely must have ac power. All lighting and much of the electronics are powered directly from the battery on 12 volts. Refrigeration, water heating, and cooking are powered by propane and are not included in this electrical estimate. There are no large motors in this estimate.

Table 9-1 is a table for organizing the information on all the various dc loads. Table 9-2 is the ac power estimate for the same small system. These tables have headings for the wattage of each appliance, the number of hours per day the appliance is used, and the number of watt-hours the device has consumed in a days time. The ac power estimate also includes a column for starting surges. These figures are then totaled to give an estimate of energy consumption.

The small system estimated in Tables 9-1 and 9-2 is optimized for efficiency. Different lifestyles

Table 9-1. Dc Energy Estimate for a Small System.

DEVICE	DEVICE WATTAGE	ON TIME PER DAY	WATT-HRS. PER DAY
LIGHTING- 2 BULBS @ 30 WATTS	60	5	300
MOTOR- SOLAR HEATING AIR BLOWER	70	1	70
COLOR TELEVISION- 10 INCH	48	5	240
STEREO- CASETTE AND FM RECEIVER	15	6	90
RADIOTELEPHONE TRANSMITTER	50	0.5	25
RADIOTELEPHONE RECEIVER	5	24	120
SMALL NI-CAD CHARGER	4	1.2	4.8
ELECTRONIC FIELD FENCER	0.3	24	7.2
INVERTER STANDBY	1	24	24
SOLDERING IRON	30	0.5	15
DC POWER CONSUMED DAILY WATT HOURS			896

will use different appliances and comsume different amounts of power. Since this hypothetical system is located far from various utilities, a radiotelephone is included in the estimate. Notice that the radiotelephone receiver is left on all the time. The usage of the radiotelephone transmitter is listed separately. This same technique can be used for all appliances which consume power at two or more rates.

In Table 9-2, notice that there is an allowance made for inverter efficiency. This table lists the energy lost within the inverter as 10 percent. This is a minimum figure, often inverters lose 20 percent to 30 percent in the dc to ac conversion process. It is essential to only use high efficiency inverters. See Chapter 8 for more information on inverters.

The total energy consumption of the hypothetical small system is the sum of both the dc and ac estimates. In this case, dc usage is 896 watt-hours per day. The ac usage is 713.35 watt-hours per day. Total energy consumption is 1,609.35 watt-hours per day. This amount is really miniscule when compared with an average consumption of around 50,000 watt-hours per day in the average American household. Information regarding the actual sizing of the inverter and battery pack is given later in this chapter.

Energy Estimate for a Large System

The second hypothetical system to be

Table 9-2. Ac Energy Estimate for a Small System.

DEVICE	STARTING SURGE	DEVICE WATTAGE	ON TIME PER DAY	WATT-HRS. PER DAY
MOTOR- VACUUM CLEANER	1500	750	0.25	187.5
COMPUTER	60	60	4	240
COMPUTER PRINTER	180	180	0.5	90
ELECTRIC DRILL	700	350	0.2	70
SEWING MACHINE	160	80	0.2	16
BLENDER	300	100	0.1	10
ELECTRIC MIXER	700	350	0.1	35
TOTAL SURGE WATTAGE ►	3,600	AC POWER CONSUMED DAILY - WATT HOURS ►		648.5
		PLUS 10% FOR INVERTER INEFFICIENCY		64.85
		AC POWER CONSUMED DAILY WATT HOURS ►		713.35

estimated is for a family of four persons, again totally without commercial power. In addition to size, this system differs greatly from the smaller system estimated in Tables 9-1 and 9-2. In the large system, all the electrical appliances are run on 120 Vac power supplied through the inverter.

Table 9-3 estimates the power needs for this large system. The system contains a refrigerator and a deep well pump. This estimate considers that each of the four family members will use 100 watts of lighting an average of 5 hours daily. Again, different lifestyles will consume different amounts of power. This estimate is offered as an example of an alternative energy system sized to meet the needs of an energy conscious family. While the system may not include all the modern conveniences, it does contain the major electrical appliances normally present in a country homestead.

Total energy consumption in this estimate is 13,257.75 watt-hours per day. This is still far below the national average power consumption. In order to meet this estimate, the family must be conservative with power usage. The actual estimation of the inverter's and battery pack's capacity is detailed later.

HOW TO SIZE THE INVERTER

In order to be efficient the inverter must not be sized too large for a given system. If the inverter is too small, then it will not operate all the equipment. Proper proportion is essential in all the components in an alternative energy system, and especially in the inverter.

In sizing the inverter's capacity, two factors must be considered: one is the average wattage consumed at the same time (average power consumption), and the other is the surge wattage (peak power consumption).

Sizing for Peak Power Consumption

The inverter must at least be sized to accommodate the largest starting surge expected in the system. For example in the small system (Table 9-2), the largest starting surge is 1,500 watts for the vacuum cleaner. In the large system (Table 9-3), the largest starting surge is 6,000 watts for the deep well pump motor. These figures determine the minimum surge wattage that the inverter must be capable of.

If in the large system it is desired to have the refrigerator and the well pump on their automatic controls, then the minimum starting surge must be rated higher. Assume that the refrigerator and the well pump start simultaneously. The starting surge would then be 9,000 watts.

The inverter should be sized to handle the starting wattage of all appliances which are able to turn themselves on. If the appliances are operated manually and if the user is careful not to start more than one device at a time, then the inverter can be sized much smaller. If energy management is practiced, the inverter need only be able to start the device with the largest surge consumption.

Sizing for Average Power Consumption

The average power consumption will be far less than the peak (surge) power consumption. Exactly how much less is determined by how many appliances are running at the same time. This is more difficult to estimate in larger families. The greater the number of people in the household, the more appliances will be in use at the same time.

It is necessary to determine how many devices are drawing energy at the same time. It is best to assume that any device which can turn itself on is operating. This estimate should include a full complement of lighting, communication, and essential uses. This information is highly personalized. The more energy management is practiced, the less power consumption. The actual estimation of the large system is specified in Table 9-3.

Inverter for a Small System

In the small system (Table 9-2), the maximum ac consumed from the inverter is 648.5 watts. A 1,000 watt inverter will handle the job, even if everything is running at the same time. The peak surge in the small system is 1,500 watts. At 1,000 watt inverter will deliver about 3,000 watts of surge power.

Table 9-3. Energy Estimate for a Large System.

DEVICE	STARTING SURGE	DEVICE WATTAGE	ON TIME PER DAY	WATT-HRS. PER DAY
WELL PUMP-2 HP.	6,000.	2,000.	1	2,000.
REFRIGERATOR	3,000.	1,000.	4	4,000.
LIGHTING	400.	400.	5	2,000.
TV- 19 INCH	200.	200.	6	1,200.
STEREO	50.	50.	5	250.
WASH MACHINE	3,000.	1,000.	0.25	250.
CLOTHES DRYER	3,000.	1,000.	0.25	250.
POWER TOOL	1,500.	750.	0.25	187.5
VACUUM	1,500.	750.	0.25	187.5
COMPUTER	60.	60.	5	300.
PRINTER	180.	180.	0.5	90.
MIXER	700.	350.	0.25	87.5
SEWING MACHINE	160.	80.	0.5	40.
VIDEO CASSETTE	150.	150.	4	600.
HAIR DRYER	750.	750.	0.5	375.
NI-CAD CHARGER	4.	4.	2	8.
RADIOPHONE RX	8.	8.	24	192.
RADIOPHONE TX	70.	70.	0.5	35.
TOTAL SURGE WATTAGE ➡	**20,732**	**AC POWER CONSUMED DAILY- WATT-HOURS ➡**		**12,052.5**
		PLUS _10_ % FOR INVERTER INEFFICIENCY		1,205.25
		AC POWER CONSUMED DAILY WATT-HOURS ➡		**13,257.75**

It is not cost effective to purchase an inverter that is under 1 kilowatt. Cost per watt on inverters goes up radically as the inverter gets smaller. The 1,000 watt inverter is the smallest size to consider for alternative energy power service. This is true for all small systems—homes, recreational vehicles, boats, or whatever. In the small system, a one kilowatt inverter has enough extra power available for future system expansion.

It's highly unlikely that all the ac appliances in the small system will be running at the same time. Most of the time the inverter will only be putting out about 200 watts or less. It is critical that the inverter have very high efficiency when it is lightly loaded. Many inverters have their efficiency rated at full output. Check to see the efficiency is also high at levels of around 30 percent to 50 percent of rated output.

Inverter for a Large System

The total wattage of all the appliances listed in the large system (Table 9-3) is around 8,800 watts. This is with everything turned on at the same time. The largest single surge needed is the well pump at 6,000 watts. Let's assume that we wish to leave the refrigerator and the well pump on automatic control. The surge wattage requirement is now up to 9,0000 watts.

It is now necessary to list all the equipment which must be able to run at the same time. The refrigerator, well pump, lighting, TV, stereo, computer, and radiotelephone are chosen for this list. The items on this list have a total wattage of around 3,800 watts.

A 5,000-watt inverter with a surge rating of 10,000 watts or greater will meet the basic need, especially if the well pump is placed under manual control. If the family insists on running everything at the same time, then the inverter should be sized at about 12,000 watts. The larger inverter (12 kilowatt) will be more expensive to purchase and will not be as efficient at an output of around 4 to 5 kilowatts.

The 5 kilowatt inverter can happily meet the need at much less cost if the family will practice energy management and not run too many large appliances at the same time. Put the well pump on manual control and use it only during periods when power consumption is low. This will further reduce the peak consumption. The use of a motorized generator to handle peak output periods will greatly increase system efficiency.

Inverter Input Voltage

Inverters of less than 2,000 watts output are usually powered by 12-Vdc inputs. Inverters of around 2,000 watts to 3,000 watts are powered by 24 Vdc. Inverters of 5,000 watts are powered by 48 Vdc. Inverters over 5,000 watts are usually powered by 140 Vdc. In other words, the higher the output power of the inverter, the higher its input voltage is. In actual practice, this translates to more batteries in series configuration to obtain the higher voltages.

It is not practical to use the dc energy directly from battery packs with voltages over 12 volts. There are simply no appliances commonly available for 24 and 48 Vdc. Users of 24- and 48-volt battery packs may be tempted to tap the pack at a 12-volt level to obtain 12-volt power. This practice is absolutely not acceptable. It can lead to very premature battery wear due to imbalances within the battery pack. The element of the battery being used for 12-volt power is also a series element of the larger 24- or 48-volt battery. As such, every ampere-hour withdrawn from the 12-volt level is made unusable at the 24- or 48-volt level. The resulting differing states of charge of the batteries making up the pack leads to great difficulty in recharging the entire battery. If the entire battery pack has a voltage of 12 volts, then the dc energy may be used directly from the battery. If the battery pack is sized at 24 or 48 volts, then all usage should be through the inverter only.

In actual practice, 12-volt dc appliances are only practical in small systems with inverters of around 1,000 watts. If the inverter is larger, the battery will have a voltage of 24 volts or greater. In systems using inverters with output power ratings of over 2,000 watts, all power use should be ac through the inverter.

HOW TO SIZE THE BATTERY PACK

An estimate of energy consumption is the first factor necessary for sizing a battery pack. The second factor is the input voltage of the inverter, if one is being used. The third factor is the number of days the battery should provide energy without being recharged.

The battery should be able to power most stand-alone alternative energy systems for at least seven days. If the sole power input is either wind or solar, then the number of days of storage should be increased to at least 14 days. Actual sizing is very difficult in systems using wind or solar as the only form of input power. The amount of charging energy is highly dependent on the weather. If a motor/generator is in use in addition to wind and solar, the battery can be sized to last seven days. These figures are highly generalized. You must decide how many days of power you wish to store.

Batteries used as standby backup sources for commercial power should be sized to meet the average length of power outage. This type of backup only requires small capacity battery packs. If the pack has only 20 percent or less of its energy removed per cycle, then it is in float service. As such, it may be composed of battery types that are less expensive than the deep cycle type.

Capacity

The ampere-hour capacity required in the battery pack can be calculated by using the following formula:

$$C = \frac{(Pd)(D)(1.25)}{Vb}$$

C = Capacity of the battery pack expressed in ampere-hours
Pd = Estimated power consumption expressed in watt-hours per day
D = Number of days between battery charges expressed in days
Vb = Voltage of the battery pack expressed in volts

The factor of 1.25 in the equation compensates for the fact that the battery will only be cycled to a 20 percent state of charge before being refilled. Cycling the battery below 20 percent of its rated capacity will result in premature battery wear and reduced system efficiency. If the battery pack is cycled to other than 20 percent state of charge before refilling, the factor in the equation must be changed.

As an example, let's specify a battery pack for the small system estimated in Tables 9-1 and 9-2. The total energy requirement is for 1,609 watt-hours per day. The system voltage is 12 volts. Lets say we wish the battery to go 7 days before recharging.

$$C = \frac{(Pd)(D)(1.25)}{Vb}$$

$$C = \frac{(1609)(7)(1.25)}{12}$$

$$C = 1{,}126 \text{ ampere-hours}$$

A 12-volt battery pack with a capacity of 1,126 ampere-hours or more will meet the need for the small system. Such a battery pack will be composed of series and parallel interconnection of a number of sub-batteries, For example, consider the Trojan L-16 which is a 6-volt, 350 ampere-hour deep cycle lead acid battery. If the battery pack for the small system were composed of L-16 sub-batteries it would require eight of them. Four paralleled batteries, each composed of a series string of two batteries, would make up the battery pack. Such a pack is illustrated in Fig. 9-1. The resultant ampere-hours capacity of such a battery pack is 1,400 ampere-hours, which exceeds the estimated capacity requirement. The cost of this battery pack will be around $1,600.

As another example, let's determine the capacity of the battery pack needed for the large system estimated in Table 9-3. The energy estimate for this system is 13,258 watt-hours per day. The battery pack voltage is 48 volts to support a 5,000 watt high

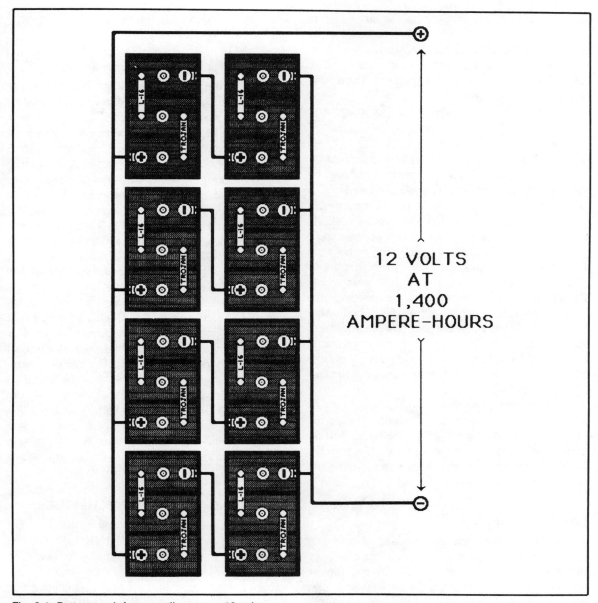

Fig. 9-1. Battery pack for a small system—12 volts.

efficiency inverter. All energy usage is 120 and 240 Vac from the inverter. Let's assume we wish to have seven days capacity in the battery pack.

$$C = \frac{(P_d)(D)(1.25)}{V_b}$$

$$C = \frac{(13{,}258)(7)(1.25)}{48}$$

$$C = \frac{2{,}417}{\text{ampere-hours}}$$

The 48-volt battery pack for the large system should have a capacity of 2,417 ampere-hours or more. If the battery pack for this system were composed of Trojan L-16s batteries, it would require 56 of them. Seven paralleled batteries, each composed of eight batteries in series, would make up a 48-volt pack with a capacity of 2,450 ampere-hours. This is a very large battery pack. The cost will be around $12,000.

The size and cost of the battery specified for the large system could be substantially reduced by energy management. For example, an 120/240 Vac powerplant can supply all of the power during peak periods of consumption. The well pump, washing machine, and other high energy consumers can be powered directly from the ac generator, rather than from the battery pack. To accomplish this the ac generator need only be run about 1 hour a day. Peak consumption is drawn from the generator and the battery pack is charged during this period rather than being heavily discharged. The capability of running the entire load from the batteries is still present, but used only when necessary. Practices such as this can cut the capacity needed in the battery pack in half.

The estimation process used here assumes that all the energy being consumed during the period of time specified is supplied by the battery pack. If there are other energy sources available, then the battery pack may be sized with less capacity and still last the same period of time.

Lead-Acid Batteries

If the battery pack is composed of lead-acid cells, then the capacity of the pack must be compensated for temperature. See Chapter 2 for the data on temperature vs. capacity for lead-acid cells. The effect of temperature on the battery pack's capacity must be allowed for in sizing the pack.

Battery capacity is rated at room temperature. If the battery pack is to be efficient and long lived, its temperature must not be too cold. Keep lead-acid cells at room temperature in a well ventilated environment. Failure to do this will result in lessened battery capacity, higher operating cost, and decreased battery longevity.

Power Output of the Charge Source

The battery pack's capacity determines the minimum output current necessary in the charging source. In stand-along applications, the power source must be capable of delivering at least a C/20 rate of charge to the battery. If the charge source is not capable of this rate, the batteries will not be cycled properly. It will not be possible to equalize the cells that make up the battery pack.

For example, consider the small system estimated in Tables 9-1 and 9-2. The battery pack is specified at a capacity of 1,400 ampere-hours at 12 volts. A C/20 rate for this battery pack is 70 amperes. The power source must be able to deliver at least 70 amperes at about 15 volts to the battery pack. This amounts to some 1050 watts. Such a power source would take about 24 hours to fill the battery pack. A five horsepower lawnmower engine driving a large automobile alternator is ideal as a primary power source for the small system.

In the large system (Table 9-3) the specified battery pack has a capacity of 2,450 ampere-hours at 48 volts. A C/20 rate for this battery pack is 122.5 amperes. A charge output of 122.5 amperes at a voltage around 60 volts is 7350 watts. This system can be run from an 8,000 watt ac motorized generator.

ENERGY MANAGEMENT TECHNIQUES

The premier technique of energy management is really very simple. IF YOU ARE NOT USING IT, THEN TURN IT OFF. This technique even works for those using commercial power. For those using energy stored in batteries, it is critical due to the even higher cost of the energy. Alternative energy systems must be expensively oversized to accommodate lightbulbs burning in empty rooms, television sets entertaining no one, and other forms of thoughtless waste.

The timing of charging and consumption are also factors which can be controlled to our advantage. This is especially true if a motorized powerplant is used in the system. Selection of efficient appliances can improve system efficiency.

Proper Timing of Peak Consumption

The proper timing of peak consumption is affected by whether or not the charge source is producing energy at the time. Energy supplied from the power source should be used directly from the source whenever it is available. If the charge source is supplying energy, then it should be loaded to its maximum. This is the time to pump the well, wash the clothes or whatever. The energy used from the source is 30 to 40 percent more efficient than energy retrieved from battery storage. This is due to the combined inherent inefficiencies in chemical batteries and inverters.

If the charge source is not supplying energy, then energy consumption should be spaced out in time to reduce peak consumption. If the battery pack is supplying the energy, it is more efficient not to turn on everything at once. If a period of time occurs when large consumption is necessary, then it is time to turn on the motorized generator.

Proper Time for Charging

The right time to charge the battery is any time that it is necessary to consume large amounts of power. Obviously, if the battery is below 20 percent state of charge it is also time to charge. Even if the battery pack is full, it is more efficient to turn on the generator during times of high power consumption.

In general, if a motorized generator is used as a power source, it should be run at night. The generation then coincides with a period of higher usage. The ambient temperature is lower at night resulting in higher efficiency in the motorized generator.

Usage of a Motorized Ac Powerplant

At the current state of the art, alternative energy sources such as wind and solar are not cost effective as stand-alone power sources. In terms of output wattage, a wind system costs about four times as much per peak watt as a motorized powerplant. Solar cells run about ten times the cost per peak watt. Add to this the fact that the energy generated from wind and solar sources is inherently unreliable. While wind and solar can contribute a significant amount of energy to a system, it is usually necessary and cost effective to have another source available.

The motorized powerplant has several advantages as a backup to wind or solar energy sources. In terms of power output in watts per hour, the motorized powerplant is very cheap. Costs are less than $1 per watt. Compare this with about $4 per watt for wind sources, and about $10 per watt for solar sources. Motorized powerplants for home use are available in sizes from 300 to 20,000 watts. This energy source is reliable; we can use it whenever we wish to start the motorized powerplant. The energy will continue as long as the plant is running. This degree of control over energy generation is simply not available from wind and solar sources.

It is true that the energy generated by a wind or solar system costs virtually nothing once the initial investment is made. A motorized powerplant, however, requires either gasoline or diesel fuel, both expensive items. The following estimates assume that the motorized powerplant is the only source of power. A small system, such as the one in Tables 9-1 and 9-2, will burn approximately 20 gallons of fuel a month. This amounts to about $300 yearly for fuel. The motorized small system produces power for around $.50 per kilowatt-hour. The large system (Table 9-3) will burn around 70 gallons of fuel monthly, or about $1,000 yearly. This amounts to around $.25 per kilowatt-hour.

Even if fuel and engine maintenance is added to the total bill, the motorized powerplant is still much more cost effective. In the small system, payback time for the initial investment in wind power would be ten years; payback time for solar is 30 years. In the large system, payback time is 26 years for wind and over 70 years for solar.

Until the cost per watt of wind and solar

sources decreases, they are not competitive with motorized powerplants. It is unlikely that wind sources will decrease radically in cost. There is simply too much high-tech manufactured material in a wind system. Solar cells, however, may be become much cheaper in the near future. A drop to under $2 per watt would make the solar cell cost effective for mass usage.

A motorized generator, if used with battery storage, is the most cost effective source of non-commercially produced power now available. The motor/generator and batteries should be considered as a stand alone startup system. Wind and solar sources may be added later.

APPLIANCES

An appliance is any device which uses electricity as a power source. It may be a refrigerator or it may be a flashlight. Appliances which are powered by batteries must be the most efficient type available. Energy used from a battery costs 5 to 50 times that of commercially produced power. The higher cost of battery-stored energy makes purchasing the more efficient appliances more cost effective. This is true even if the energy efficient appliance costs much more than other less efficient types.

For small alternative energy systems using 12 volts as a battery pack voltage, there are many 12 Vdc appliances available. This is not true for larger (over 2000 watts) systems. Larger systems have battery pack voltages of 24 volts and more and are limited to standard 120 Vac appliances powered through the inverter. These standard 120 Vac appliances are designed to sell in a highly competitive mass market. As such, they are made to sell cheaply rather than to be energy efficient. It is worth your time to make sure that the appliance you are considering for battery power is the most efficient type available.

Lighting

Lights are commonly powered by battery-stored energy. One of the very first successful applications of batteries was in flashlights. Batteries now power emergency lighting, boat lighting, RV lighting, and lighting in alternative energy households. Many of these lighting applications use direct current, low voltage energy directly from the battery. In most cases, standard 120 Vac lighting is powered from batteries via an inverter. Whatever type of power is used, there are techniques that can reduce the operating cost of the lighting.

It is common to use area lighting and indirect lighting techniques in homes powered by commercially produced electrical power. These lighting techniques are wasteful of electrical power. Areas which do not require illumination are illuminated. The usage of area lighting should be held to an absolute minimum in battery powered systems. Lighting should be of the spot type and placed as close as possible to the place where it is needed. This is true for both 12 Vdc and 120 Vac lighting.

When using battery powered lighting, it is much more efficient to place many small spotlights around a room than it is to use one large light for illuminating the entire room. Some of the small lights may be switched off when not in use. The spotlight needs less wattage to do the job because it is located close to where it is needed. A 40-watt bulb on the desk will better illuminate the desk than a 200-watt bulb on the ceiling.

Twelve volt dc lighting usage directly from battery storage is a very efficient method. There is a wide variety of incandescent lightbulbs commercially available due to their usage in the automobile. These small bulbs use between 10 and 40 watts of power. They are very suitable for spotlighting. In fact, the commercially available high intensity desk lamps use automotive lightbulbs. There is a transformer in the lamp's base which feeds the automotive lightbulb 12 volts.

The 12 Vdc incandescent lightbulb is actually more efficient than its 120 Vac relative. The increased electron flow through the 12-volt bulb's thicker filament is more efficient in converting the electrical energy into light. More of the energy consumed by the 12-volt bulb is radiated as light and less is wasted as heat. For those with 12 Vdc systems, the automotive lightbulb is the most efficient choice for incandescent illumination. The

12-volt lighting must be connected to the battery in the proper manner or excessive power will be lost in the wiring.

Fluorescent lights are commercially available with 12-volt power inputs. These lights are super efficient. Each light contains a micro-inverter which makes high voltage power for the fluorescent tube. In the cheaper models, this micro-inverter can be very noisy. It can interfere with radios and television receivers. Only high quality lights should be considered. They cost about $1.50 per watt and are available in sizes around 20 to 40 watts. The fluorescent light can substantially reduce lighting power consumption.

In larger systems, all the lighting is powered by 120 Vac supplied from the inverter. The standard incandescent lightbulb works well on inverter power. Try to use the minimum wattage bulb that will do the job. This is especially important for lights that spend long periods of time on. It is prudent to use the newer *energy saver* types of incandescent bulbs. These bulbs give more light from their input power consumption. The efficiency picture of the incandescent lightbulb is really grim. The average 120 Vac lightbulb wastes over 90 percent of its input energy as heat.

Most types of 120 Vac fluorescent lighting will work on inverter power. The fluorescent light is much more efficient than the incandescent type. There is, however, some limitation due to incompatibility between some types of inverters and some fluorescent lights. The inverter supplies a modified sine wave form of power to the fluorescent light. This form of power can cause an annoying buzzing sound to be emitted from the light. The inverter's power is hard on all types of fluorescent lighting ballasts. The ballasts tend to wear out many times quicker on inverter produced power. Experimentation is necessary with the particular inverter and fluorescent light in order to determine compatability. In general, do not use fluorescent lighting with square wave inverters. High quality modified sine wave inverters can be used. The newer type of rapid start fluorescent is better suited for inverter power than the older types.

Lighting, regardless of its form, is a major consumer of electricity. It is not that it actually has such high consumption, but that it spends so much of the time on. Even small loads can add up if they are present for long periods of time. Usage of these techniques can reduce the power consumption of lighting while still providing all the light that is needed.

Electric Motors

Electric motors are large consumers of power. The jobs they perform are usually rated in the hundreds of watts. As such large consumers, it is very important that they are as efficient as possible and use as little as possible.

The electric motors drive refrigerators, freezers, pumps, fans, power tools, vacuum cleaners, and a myriad of kitchen appliances. The type of motor in each of these appliances is important to the users of battery-stored power. Not all motors are equal. Some types are more efficient than others.

Refrigeration

Refrigerators and freezers are available for many different types of input power. Some of the types are 120 Vac motor-driven, 12 Vdc motor-driven, 12 Vdc solid state, and propane. Of the motor driven types, the 120 Vac models are more efficient. This is due to the losses involved in the heavy current comsumption of low voltage motors. The 12 Vdc solid state refrigeration is a new development. It uses semiconductor junctions to transfer heat. This type is very efficient for small amounts of cooling. Hopefully, larger units will become available as this form of cooling technology is further developed.

Refrigerators and freezers now are sold with energy consumption estimates displayed on them. These should be considered at the time of purchase. The thickness of the refrigerator's insulation is a factor. More insulation means less energy consumption. Chest refrigerators and freezers which open at the top are more efficient than ones with vertical doors which allow the cold air to escape upon opening. The refrigerator may be located in a cool place, away from heat sources. This will decrease the amount of power required.

Simple techniques such as filling a partially filled refrigerator with bottles can reduce power consumption. Don't stand there with the door open gazing at the contents. Many techniques involved in energy management are simple common sense. The higher cost of battery-stored energy makes it essential to put this common sense into daily practice.

Pumps

Pumps are large consumers of power that are often essential. In many cases, a motorized pump can do the job without resorting to electrical use. In the case of deep well pumps, there are no alternatives; electricity must be used.

The motors in well pumps are usually powered by 240 Vac. They are usually between 1 and 2 horsepower. Only the larger inverters are capable of 240 Vac output. If an electric well pump is used, then a storage tank should be provided. Using the water directly from the well will increase the number of times the pump will start daily. The storage tank allows the pump to run continually, then shut off for a longer period of time. This means the pump may be controlled manually. It may also be powered by a motorized generator started especially for this purpose. This takes much of the load off the inverter and the battery pack, resulting in higher energy efficiency.

Electronics

Here is where battery power really begins to shine. Electronic devices are naturals for battery power. The transistor has made possible an entire new world of electronic devices. The transistor is basically a low voltage, current operated device. It thrives on battery stored power. All these small portable electronic marvels we use are filled with transistor junctions and batteries. If transistor devices are operated on standard 120 Vac, then a power supply must be used to supply the transistors with low voltage dc power. This is the same type of power that the battery stores. The users of battery stored energy actually have a much wider choice of electronics than does the consumer who only has commercial 120 Vac power available. Many electronic devices have provisions for low voltage power inputs.

One of the first electronic devices to be battery powered was the radio. Radio receivers consume very little power. In portable equipment operated by primary cells, it is desirable to use the larger sized cells, such as C or D. The energy stored in the larger cells is cheaper per watt-hour than in the smaller sized cells. Refer to Chapter 5 for information regarding primary (nonrechargeable) batteries.

All types of radio receivers and transmitters are available with input powers of 12 Vdc, for service in automobiles. These mobile units can be used unmodified in 12-volt battery systems. The average automobile FM stereo radio is actually a better deal than the home models. The receivers are more sensitive to compensate for mobile operation. These FM radios are mechanically more rugged and generally cheaper. The additional cost of the home models is due to the larger fancy case and the additional power supply to make the type of energy that battery users already have.

Audio cassette tape equipment is available in 12 Vdc models. Record turntables are usually restricted to 120 Vac via an inverter. If 120 Vac audio gear is powered from batteries via the inverter, then care must be taken in selecting efficient equipment. Modified sine wave inverter power is acceptable to most turntables and stereos.

Television sets are commonly made with 12 Vdc inputs, as well as 120 Vac inputs. The power consumption of a television set goes up in proportion with the surface area of the picture tube's screen. A 10-inch TV will consume about two times the power of a 7-inch TV. Statistics tell us that the average American family has a television set on in excess of 6 hours daily. Efficiency is important in appliances which spend so much time operating. Keep the screen size as small as is pleasing. Video equipment is available in portable battery powered models. These types are made to be more efficient than the standard 120 Vac only models. Portable equipment, even when powered from the inverter, is generally more efficient to use.

Computers will run quite well on batteries via the inverter. In fact, the computer used to write this book was powered completely from batteries via an inverter. Computer equipment constructed for 120 Vac operation usually has a tremendous filtration capacity built into the power supply. As such, it runs well on the modified sine wave inverters. This is also true for printers and peripherals. One advantage of computer use on batteries is the quality of the power. The batteries via the inverter provide a much more stable and constant form of power than can be purchased from the power company. Computers powered by batteries are not affected by brown-outs and black-outs. There is less noise and other forms of electrical trash on line with homemade power. It is an ideal source of clean continuous computer power.

In general, just about any electronic device that can be plugged into standard housepower can also be run on battery power through an inverter. The users of battery stored power need no longer to be content with electronics that have low voltage dc power inputs. The modern power inverters will run most forms of consumer electronics. This is true of even the most sophisticated electronics, such as satellite television receivers.

USING ENERGY TO REDUCE MANAGEMENT SYSTEM COST

Proper management of energy production, storage, and usage is the key to successful battery operation. Energy management can cut the initial investment in equipment to a minimum by reducing average and peak energy consumption. Realistic estimation of the amount of energy needed is the first and most difficult step. Once the estimate of power consumption is completed, the rest of the specification process is straightforward and simple. Everything is determined by the estimate of power consumption.

It is common to revise the estimate of power consumption once the entire specification process is completed. Along with the specification of battery capacity, inverter capacity, and size/mode of power supply comes a cost estimate—the proverbial bottom line. This initial cost estimate is often intimidating to those holding the opinion that alternative energy provides *free* power. It is then time to revise the power consumption estimate. It may be possible to do without the electric dishwasher, or whatever. It may be possible to sell inefficient appliances and replace them with more efficient units. The entire specification process is then repeated, generating new initial cost figures. This process of re-estimation and specification is carried on until the initial cost estimate is acceptable.

System Cost Estimates

System cost estimation includes several different kinds of estimates. There is an initial cost estimate, an operating cost estimate, and an average cost per year estimate usually based on a 10-year period. These estimates generate a figure for energy cost that is expressed in dollars per kilowatt-hour. These different forms of estimation are necessary to ensure that all factors are considered.

The initial cost estimate covers the purchase of all the hardware in the system. This hardware estimate includes the costs of the batteries, inverter, power source, and all hardware necessary to get the system interconnected and working. Any labor costs should be included here. Also allow for transportation of the system's components to the site of the installation. Batteries are very heavy objects. They are corrosive and must be shipped by motor freight. Shipping costs can be expensive and must be included in the initial cost estimate.

Operating cost estimates are usually made for a yearly period. In motorized systems include the following items—fuel, oil, and any routine engine maintenance. In wind systems the operating expenses of the actual wind machine are low—brushes, bearings, and periodically a new propeller. Tower maintenance, however, is expensive and must be included in this operating cost estimate. Solar systems are virtually maintenance-free and have no operating costs.

Alternative energy systems are usually amortized over a 10-year period. This is done because 10 years is the lifespan of the average battery pack.

The average cost per year estimate includes all hardware and all operating expenses over a 10-year period. This figure is then averaged to obtain the yearly average cost. The longevity of system components other than the batteries is an important factor in the average cost per year figure. In motor driven systems, the longevity of the engine is a factor. Inexpensive small gasoline engines are only good for around 1,000 hours of operation and will wear out in about a year. Higher quality gas engines, such as those made by Honda, will run for over 3,000 hours, or about 3.5 years before replacement. Small diesel engines commonly run over 20,000 hours before rebuilding. Initial investment is higher for engines with greater longevity. A wind system should have at least 10 years longevity. The longevity of solar cells is in excess of 10 years.

All the above estimates, along with the estimated power consumption, finally generate a figure which expresses energy cost in terms of dollars per kilowatt-hour. This is the same basis on which the commercial power companies sell their electricity. This figure is averaged over the 10-year period of estimation.

Cost Estimates for a Small System

As an example, consider the small system detailed in Tables 9-1 and 9-2. The cost estimates for this small system are presented in three forms, each using a different power source. The different sources are motorized dc alternator, wind machine, and solar cells. Each power source is evaluated as the sole source of power. The form organizes the cost information so that it can be readily compared with other systems. All cost figures are approximate and referenced to 1985 U.S. dollars.

Table 9-4 gives the information regarding the small system powered by a motorized dc alternator. Initial hardware cost is $3,450. Yearly operating cost is $320. Average annual cost over a 10-year period is $745 yearly. This estimate includes wearing out three engines and one alternator during the 10-year period. The average cost of the energy used over a 10-year period is $1.27 per kilowatt-hour. Figure 9-2 shows the cost distribution for a small motorized system over a 10-year period.

Table 9-5 shows the cost estimate for the small system using a wind power source. In terms of energy consumption, the system is essentially the same as the one in Table 9-4. The battery pack, however, is twice as large to accommodate the intermittent nature of wind power. The wattage of the wind source is increased for the same reason and to accommodate the larger battery pack. This estimate assumes that there is enough wind on site to effectively drive the generator. Initial hardware cost is $12,400. Average cost per year over 10 years is $1,340. The cost of the energy per kilowatt-hour is $2.28. Figure 9-3 shows the cost distribution of a small wind powered system.

Table 9-6 is the cost estimate for the same system, only using photovoltaic solar cells for a power source. Once again, the battery capacity and source wattage must be increased to compensate for the intermittent output of solar cells. Initial cost is $26,900. Average cost per year over 10 years is $2,690. The energy costs about $4.58 per kilowatt-hour to generate and use. Figure 9-4 gives the cost distribution for a small solar powered system.

As can be seen from these estimates, the motorized power source is by far the most cost effective. Even in a small system, the motorized power source costs about two times less than a wind power source. The motor beats solar by a factor of about 3.5 times. This is in part due to the higher initial costs of wind and solar sources. The intermittent nature of wind and solar sources makes it necessary to oversize the battery and thereby the wattage of the power source. All these factors increase the cost per kilowatt-hour of the electricity.

Cost Estimates for a Large System

The next series of three cost estimates are for the large system listed in Table 9-3. Once again, three different types of power sources are considered: motorized, wind, and solar. Each power source is considered as the only energy input to the system.

Table 9-7 gives the information for the motorized system. The powerplant is a 12.5 horsepower single cylinder diesel engine driving an 8 kilowatt ac alternator. Output from the powerplant and the

Table 9-4. Cost Estimate for a Small System Using a Motor.

DC POWER CONSUMPTION in W-hrs./day	896
AC POWER CONSUMPTION in W-hrs./day	713
TOTAL POWER CONSUMPTION in W-hrs./day	1,609
NUMBER OF DAYS OF BATTERY STORAGE	7
BATTERY CAPACITY in AMPERE-HOURS	1,400
BATTERY PACK VOLTAGE	12
ESTIMATED BATTERY COST -- 8 X Trojan L-16W	1,600
ESTIMATED INVERTER WATTAGE	1,200
ESTIMATED INVERTER COST -- Heart HF12-1200X	1,200
ESTIMATED COST: INVERTER & BATTERIES	2,800
POWER SOURCE -- 5hp. Motorized DC Charger	
POWER SOURCE WATTAGE	1,500
ESTIMATED POWER SOURCE COST	650
ESTIMATED OPERATING COST PER YEAR	320
ESTIMATED SOURCE LONGEVITY IN YEARS	3.5
ESTIMATED INITIAL HARDWARE COST	3,450
ESTIMATED COST PER YEAR OVER A 10 YEAR PERIOD	745
ESTIMATED ENERGY COST IN $ PER KILOWATT-HOUR	1.27

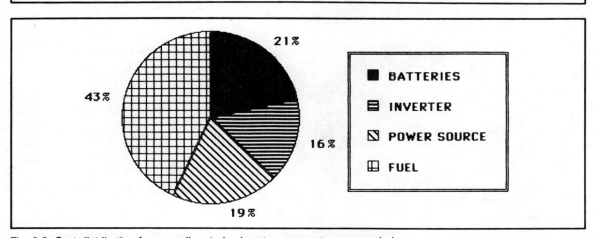

Fig. 9-2. Cost distribution for a small motorized system over a ten-year period.

Table 9-5. Cost Estimate for a Small System Using Wind Power.

DC POWER CONSUMPTION in W-hrs./day	896
AC POWER CONSUMPTION in W-hrs./day	713
TOTAL POWER CONSUMPTION in W-hrs./day	1,609
NUMBER OF DAYS OF BATTERY STORAGE	14
BATTERY CAPACITY in AMPERE-HOURS	2,800
BATTERY PACK VOLTAGE	12
ESTIMATED BATTERY COST -- 16 X Trojan L-16W	3,200
ESTIMATED INVERTER WATTAGE	1,200
ESTIMATED INVERTER COST -- Heart HF12-1200X	1,200
ESTIMATED COST: INVERTER & BATTERIES	4,400
POWER SOURCE -- Wind Powered Generator	
POWER SOURCE WATTAGE	2,000
ESTIMATED POWER SOURCE COST	8,000
ESTIMATED OPERATING COST PER YEAR	100
ESTIMATED SOURCE LONGEVITY IN YEARS	10
ESTIMATED INITIAL HARDWARE COST	12,400
ESTIMATED COST PER YEAR OVER A 10 YEAR PERIOD	1,340
ESTIMATED ENERGY COST IN $ PER KILOWATT-HOUR	2.28

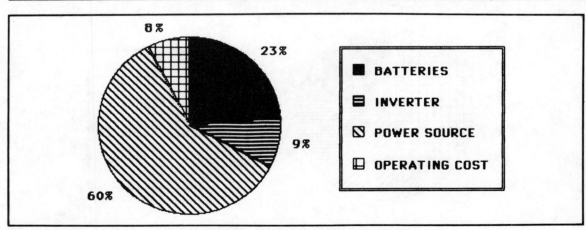

Fig. 9-3. Cost distribution for a small wind powered system over a ten-year period.

Table 9-6. Cost Estimate for a Small System Using Solar Power.

DC POWER CONSUMPTION in W-hrs./day	896
AC POWER CONSUMPTION in W-hrs./day	713
TOTAL POWER CONSUMPTION in W-hrs./day	1,609
NUMBER OF DAYS OF BATTERY STORAGE	14
BATTERY CAPACITY in AMPERE-HOURS	2,800
BATTERY PACK VOLTAGE	12
ESTIMATED BATTERY COST -- 16 X Trojan L-16W	3,200
ESTIMATED INVERTER WATTAGE	1,200
ESTIMATED INVERTER COST -- Heart HF12-1200X	1,200
ESTIMATED COST: INVERTER & BATTERIES	4,400
POWER SOURCE -- Photovoltaic Solar Cells	
POWER SOURCE WATTAGE	2,250
ESTIMATED POWER SOURCE COST	22,500
ESTIMATED OPERATING COST PER YEAR	-0-
ESTIMATED SOURCE LONGEVITY IN YEARS	10
ESTIMATED INITIAL HARDWARE COST	26,900
ESTIMATED COST PER YEAR OVER A 10 YEAR PERIOD	2,690
ESTIMATED ENERGY COST IN $ PER KILOWATT-HOUR	4.58

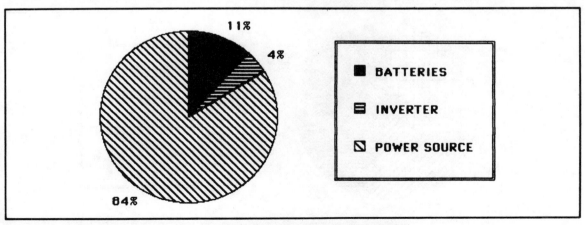

Fig. 9-4. Cost distribution for a small solar powered system over a ten-year period.

Table 9-7. Cost Estimate for a Large System Using a Motor.

DC POWER CONSUMPTION in W-hrs./day	-0-	
AC POWER CONSUMPTION in W-hrs./day	13,258	
TOTAL POWER CONSUMPTION in W-hrs./day	13,258	
NUMBER OF DAYS OF BATTERY STORAGE	7	
BATTERY CAPACITY in AMPERE-HOURS	2,450	
BATTERY PACK VOLTAGE	48	
ESTIMATED BATTERY COST --56 X Trojan L-16W		11,200
ESTIMATED INVERTER WATTAGE	5,000	
ESTIMATED INVERTER COST -- Heart HF48-5000X		3,700
ESTIMATED COST: INVERTER & BATTERIES		14,900
POWER SOURCE -- 12.5 hp. Diesel Generator		
POWER SOURCE WATTAGE	8,000	
ESTIMATED POWER SOURCE COST		6,000
ESTIMATED OPERATING COST PER YEAR	1,100	
ESTIMATED SOURCE LONGEVITY IN YEARS	10	
ESTIMATED INITIAL HARDWARE COST		20,900
ESTIMATED COST PER YEAR OVER A 10 YEAR PERIOD		3,190
ESTIMATED ENERGY COST IN $ PER KILOWATT-HOUR		0.66

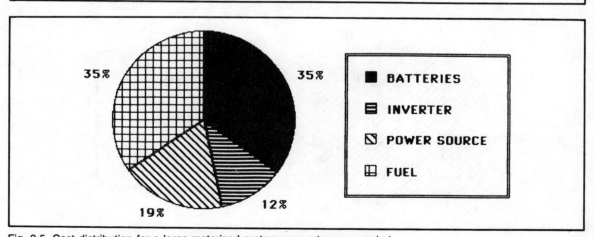

Fig. 9-5. Cost distribution for a large motorized system over a ten-year period.

Table 9-8. Cost Estimate for a Large System Using Wind Power.

DC POWER CONSUMPTION in W-hrs./day	-0-
AC POWER CONSUMPTION in W-hrs./day	13,258
TOTAL POWER CONSUMPTION in W-hrs./day	13,258
NUMBER OF DAYS OF BATTERY STORAGE	14
BATTERY CAPACITY in AMPERE-HOURS	4,900
BATTERY PACK VOLTAGE	48
ESTIMATED BATTERY COST -- 112 X Trojan L-16W	22,400
ESTIMATED INVERTER WATTAGE	5,000
ESTIMATED INVERTER COST -- Heart HF48-5000X	3,700
ESTIMATED COST: INVERTER & BATTERIES	26,100
POWER SOURCE -- Wind Powered Generator	
POWER SOURCE WATTAGE	14,000
ESTIMATED POWER SOURCE COST	56,000
ESTIMATED OPERATING COST PER YEAR	200
ESTIMATED SOURCE LONGEVITY IN YEARS	10
ESTIMATED INITIAL HARDWARE COST	82,100
ESTIMATED COST PER YEAR OVER A 10 YEAR PERIOD	8,410
ESTIMATED ENERGY COST IN $ PER KILOWATT-HOUR	1.73

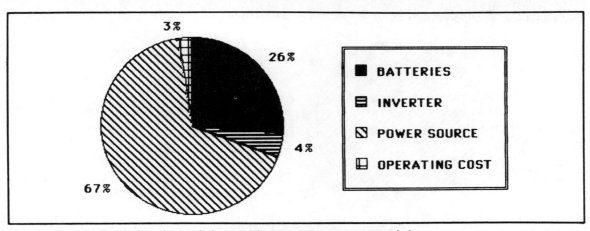

Fig. 9-6. Cost distribution for a large wind powered system over a ten-year period.

Table 9-9. Cost Estimate for a Large System Using Solar Power.

DC POWER CONSUMPTION in W-hrs./day	-0-
AC POWER CONSUMPTION in W-hrs./day	13,258
TOTAL POWER CONSUMPTION in W-hrs./day	13,258
NUMBER OF DAYS OF BATTERY STORAGE	14
BATTERY CAPACITY in AMPERE-HOURS	4,900
BATTERY PACK VOLTAGE	48
ESTIMATED BATTERY COST --112 X Trojan L-16W	22,400
ESTIMATED INVERTER WATTAGE	5,000
ESTIMATED INVERTER COST -- Heart HF48-5000X	3,700
ESTIMATED COST: INVERTER & BATTERIES	26,100
POWER SOURCE -- Photovoltaic Solar Cells	
POWER SOURCE WATTAGE	14,000
ESTIMATED POWER SOURCE COST	140,000
ESTIMATED OPERATING COST PER YEAR	-0-
ESTIMATED SOURCE LONGEVITY IN YEARS	10
ESTIMATED INITIAL HARDWARE COST	166,100
ESTIMATED COST PER YEAR OVER A 10 YEAR PERIOD	16,610
ESTIMATED ENERGY COST IN $ PER KILOWATT-HOUR	3.43

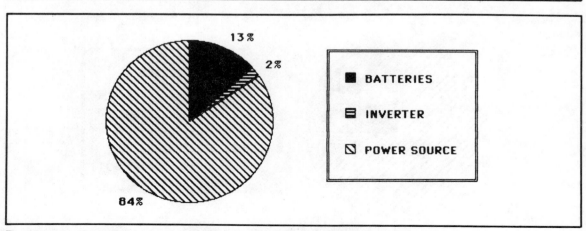

Fig. 9-7. Cost distribution for a large solar powered system over a ten-year period.

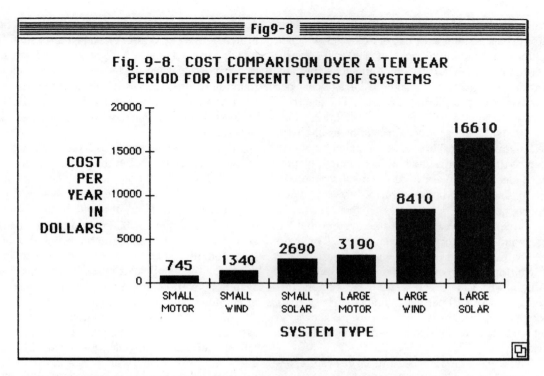

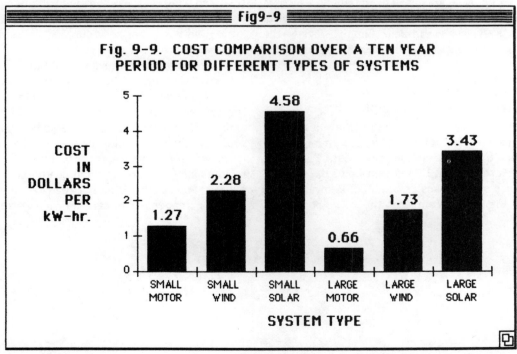

inverter is both 120 Vac and 240 Vac. The alternator drives a 7,000-watt battery charger which is included in the initial cost of $20,900. Yearly operating cost is $1,100, which includes fuel and all routine engine maintenance. Average cost per year over a 10-year period is $3,190. The electricity costs about $.66 per kilowatt-hour. This represents the cheapest electrical power of any of the examples used here. Even as such, it is still around 10 times the cost of commercially produced electrical power. Figure 9-5 shows the cost distribution for the large motorized system over a 10-year period.

Table 9-8 estimates the same system, but using a wind generator as a power source. The oversize batteries and source wattage reflect the intermittent nature of the wind source. Initial cost is high, $82,100. Over a ten-year period this averages to $8,410 per year. The electricity costs about $1.73 per kilowatt-hour. Figure 9-6 shows the cost distribution for this system over a 10-year period.

Table 9-9 estimates the same large system, only powered by solar cells. The initial investment is astronomical, over $166,000. The energy costs about $3.43 per kilowatt-hour. Figure 9-7 shows the cost distribution for the large solar system.

Cost Comparison of Different Type of Systems

The cost estimates of all the different types of systems may be compared to determine which delivers the power at the least cost. The small system uses around 1.6 kilowatt-hours daily, while the large system uses around 13.3 kilowatt-hours daily. Figure 9-8 compares the cost per year in dollars for the different systems. The cost figure is an average per year over a 10-year period. Here it is apparent that the more power used, the more expensive the system. There is no free lunch to be found here. It is interesting to note that the small solar system costs about the same as the large motorized system, which delivers about eight times as much energy.

Figure 9-9 shows a comparison between the different systems in relation to the cost per kilowatt-hour of the electricity consumed. The large motorized system produces energy the cheapest, about $.66 per kilowatt-hour, or about 10 times the cost of commercial power. This is due to the cheaper cost (in $/kW.-hr.) and greater efficiency of the larger engines. The most expensive energy is produced by the small solar system. The small solar system delivers energy at about 90 times the cost of commercially produced power.

An analysis of the information based on these estimates leads to the conclusion that motorized power sources are the only type to use at the present time. Perhaps this will change in the future as solar and wind sources come down in price. Those considering remote power should establish the system with a motorized source. Solar and wind sources can be added as they become cost-effective.

The cost of the batteries in all of the systems estimated is an appreciable part of the money spent on the entire system. In motorized power systems, the batteries account for between 20 and 35 percent of the expense. If the batteries are not used properly, then they will not survive the 10-year period of estimation. Improper battery cycling and maintenance will raise overall costs, resulting in electrical power that is more expensive. It is good energy management to care for your batteries.

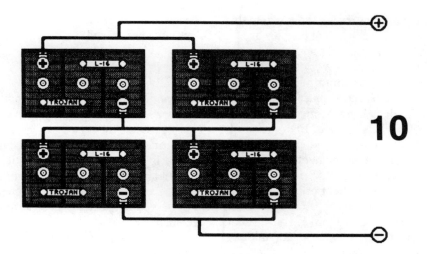

Developing Battery Technologies

This chapter discusses the newer forms of battery technology. Battery technology is not static. Much research is being carried on by business and government organizations. This research is being done to increase battery performance and lower battery cost. Some research projects work to improve currently manufactured types of batteries. Other research projects are developing entirely new types of batteries. Most of the research being conducted is oriented to two specific battery applications—electric vehicles and utility load leveling.

One of the prime motives for improving battery performance is the possibility of electric vehicles. The technology already exists to produce and market a usable electric vehicle with one exception—the battery. If a suitable battery can be developed, then the electric automobile will become a viable reality. The problems to be overcome in vehicle service are energy density, safety, and cost. Research, funded in part by the federal government, is being carried out under the Electric and Hybrid Vehicle Research, Development and Demonstration Act of 1976.

The type of battery needed to power electric vehicles must have a very high energy density. It must be able to store a large amount of power in relation to its weight. This is one of the major problems with current battery types: they are simply too heavy. A vehicle battery must store large quantities of energy and still be low in weight. Research done in this area is unlikely to benefit stationary users of battery-stored energy.

The electric vehicle battery is being designed for moderate discharge rates, around C/5, with short periodic surges of much higher rates. The battery must also be capable of accepting the intermittent service involved in motive applications. During periods of acceleration, energy is removed from the battery. During periods of deceleration, the inertia of the car is used to charge the battery. This type of intermittent service is very similar to that encountered in wind and solar powered systems.

Another type of battery being developed is designed for utility load leveling applications by commercial power companies. The idea here is to reduce peak energy generation requirements by storing energy in batteries during times of less than peak usage. New electrochemical couples are being tried to find a more efficient and cheaper battery. Some of these types are very sophisticated, employing electrolyte pumps and high temperatures. These batteries must first of all be cost effective in delivering large quantities of energy for short periods of time. Energy density and thereby weight are not considerations. These batteries will be massive permanent installations, used and maintained by experts.

The design requirements of utility load leveling batteries are very different than those of alternative energy service. Since storage is for a short period of time, rates of self-discharge are unimportant. The technical complexity of the load leveling types makes the initial expense too great for consumer use. It is unlikely that batteries developed for utility load leveling will have consumer potential, other than for very large emergency short-term power backup.

All the new batteries discussed in this chapter are purely experimental. You can't buy these batteries now at your local battery shop. We are looking between 7 to 20 years into the future. Early indications are that it may be possible to produce batteries with much greater efficiencies and lower prices. How much these new batteries will cost is ultimately determined by consumer demand. All of these types will be very expensive until high demand justifies automated mass production.

NICKEL-ZINC

The nickel-zinc cell is very similar to the Edison cell. It represents an attempt to rectify some of the deficiencies present in the nickel-iron cell. The nickel-zinc cell has potentially higher energy density than either the Edison or lead-acid cell. Its active materials are cheap and commonly available. These combined factors make nickel-zinc cells a potential candidate for electric vehicle service.

Current prototypes of the nickel-zinc cells exhibit very poor cycle life. Research is centered on prolonging the cell's lifetime and reducing its cost. It is unlikely that the nickel-zinc cell will be suitable for alternative energy service. This is due to its relatively short lifetime in comparison to already existing types of cells.

Projected Performance

The nickel-zinc cell is being optimized to perform best at discharge rates between C/5 and C/1. This represents the type of discharge service found in electric vehicular applications. Long term energy storage is not considered. The cells are designed to be emptied and refilled in less than a 24-hour period. As such, the relatively high self-discharge rate of the nickel-zinc cell is unimportant in vehicular service.

Cell voltage is somewhat higher than the nickel-iron or nickel-cadmium cells. This is a contributing factor to the nickel-zinc cells' projected higher energy density of around 90 watt-hours per kilogram. The cell voltage of the nickel-zinc electrochemical couple is around 1.6 volts. The cell voltage is relatively constant over the entire discharge cycle. The voltage of the nickel-zinc cell seems to be less affected by temperature and high rates of discharge than its near cousin the Edison cell. Figure 10-1 illustrates the relationship between cell voltage and depth of discharge for several discharge rates. Figure 10-1 assumes the cell temperature is 78° F. Temperature has an effect on both cell voltage and apparent cell capacity. Figure 10-2 gives the relationship between cell voltage and depth of discharge for various temperatures. Figure 10-2 assumes a discharge rate of C/3.

The nickel-zinc cells are being developed in capacities around 300 ampere-hours. This represents a size optimized to vehicular service. Temperature plays a large part in the actual usable capacity of the nickel-zinc cell. The optimum operating temperature is around 80° F. Figure 10-3 illustrates the effect of temperature on the nickel-zinc battery.

The nickel-zinc cell prototypes exhibit the distressing characteristic of being ruined by overcharging. Overcharging causes dendrites of zinc to grow between the cell's plates, causing an internal short circuit within the cell. This condition ruins the cell. Nickel-zinc cells are not capable of being

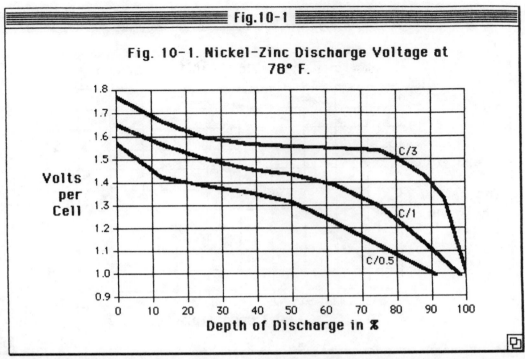

Fig. 10-1. Nickel-Zinc Discharge Voltage at 78° F.

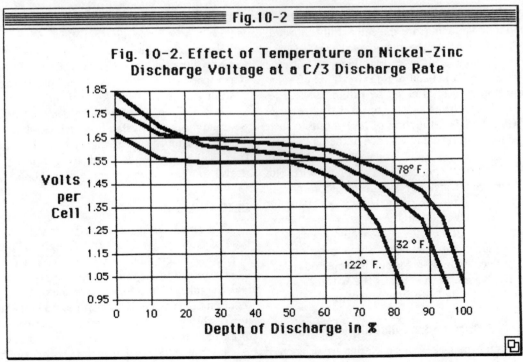

Fig. 10-2. Effect of Temperature on Nickel-Zinc Discharge Voltage at a C/3 Discharge Rate

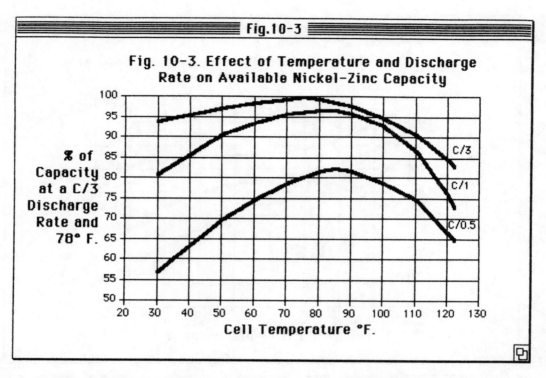

Fig. 10-3. Effect of Temperature and Discharge Rate on Available Nickel-Zinc Capacity

charged at rates exceeding C/10. Cell voltages around 2.05 volts indicate the cell is fully charged.

The efficiency of the nickel-zinc cell is between 60 to 65 percent. This is comparable with both the lead-acid and nickel-cadmium type. This efficiency figure is greatly affected by rate of discharge. The efficiency drops sharply at rates of discharge over C/1. This is a function of losses due to heat caused by the internal resistance of the nickel-zinc cell.

Cycle life is the biggest problem with the nickel-zinc cell. Currently produced cell prototypes are worn out after less than 100 cycles. It is optimistically predicted that the cycle life may be extended to over 300 cycles. This would necessitate yearly replacement in electric vehicular service. In float service, the nickel-zinc cell is projected to last about five years. If the cost of the battery can be reduced, its limited lifetime is acceptable.

Projected Cost and Availability

Costs for the nickel-zinc batteries are projected to be around $60 per kilowatt-hour. If this cost projection can be achieved, then this type of battery will be among the cheapest types available per kilowatt-hour. This estimate reflects automated mass production. Current cell prototypes are costing around $500 per kilowatt-hour to produce.

It is unlikely that the nickel-zinc batteries will be commercially available before the late 1980s. These first units will be highly optimized to electrical vehicular service and are not likely to be useful to alternative energy applications.

ZINC-CHLORINE

The zinc-chlorine cell is very different from any type of battery discussed to this point. This type of cell uses a circulating water-based electrolyte. This electrolyte is pumped to and from an electrolyte tank which is external to the cell where the reaction takes place. Energy is stored in the ionization of chlorine molecules, which are then pumped to an external tank for cooling and storage.

The zinc-chloride battery stores energy by the

ionization of a solution of zinc chloride in water. When electrical energy is applied to the cell, the zinc chloride solution dissociates into zinc metal which plates one electrode, while chlorine gas forms on the other electrode. The electrodes of the cell do not actually participate in the electrochemical reaction. The electrodes are merely sites for ion formation and are constructed of inert materials. As such, the zinc-chlorine cell has no inherent polarity. When the cell is initially charged, zinc plates the cathode and chlorine gas evolves at the anode.

As the cell is charged some of the chlorine gas evolving at the anode becomes dissolved in the water electrolyte. Some remains in the gaseous state. The electrolyte is then pumped into the external tank. Any gaseous chlorine is also transferred to the external tank by a separate gas pump. This external tank contains electrolyte held at a temperature of less than 48° F. The chlorine gas dissolved in the electrolyte forms chlorine hydrate, a pale yellow solid, as it is cooled. The heat released by the formation of the chloride hydrate is removed from the tank in order to keep the electrolyte temperature below 48° F. At temperatures above 48° F. (at atmospheric pressure) chloride hydrate becomes unstable and the chlorine once again becomes gaseous.

During the charge cycle, the zinc dissociates from the zinc chloride in the electrolyte solution. The zinc metal plates the cathode of the cell, where it remains until the cell is discharged. The amount of zinc chloride in solution in the electrolyte decreases as the cell is charged. Energy is stored by breaking apart the zinc chloride bond and storing the zinc and chlorine ions for later recombination.

During the discharge cycle, the electrolyte is pumped back into the cell for reaction with the zinc plated to the cathode. The electrolyte solution of chlorine hydrate and water is heated before introduction to the cell. This is done to return the chlorine to the gaseous state for reaction with the zinc. The amount of zinc chloride in the electrolyte increases as the battery is discharged.

The zinc-chlorine cell is being developed to meet both vehicular and load leveling type of service. This type of cell is being developed in two forms—stationary and mobile. Both types are designed to hold energy for short periods of time, less than several days. Both types use electrolyte and gas pumps and both require refrigeration of the chlorine hydrate storage tank.

The zinc-chlorine cell is capable of being totally discharged repeatedly without damage. A periodic maintenance technique is to totally discharge the battery and reverse the polarity of the electrodes. This minimizes zinc dendritic growth and equalizes the individual cells. The zinc-chloride cell is remarkable in this respect. Cell reversal would ruin most types of electrochemical cells.

Projected Performance

Cell voltage for the zinc-chlorine reaction is around two volts. It is interesting to note that the cell's rest voltage is not dependent on the cell's state of charge. It remains constant. Factors affecting cell voltage are rate of charge/discharge, electrolyte temperature, and electrolyte flow rate. Figure 10-4 illustrates cell voltage as a function of charge and discharge rates. Electrolyte temperature is assumed to be 122° F. The zinc-chlorine cell keeps a more constant output voltage over the entire discharge cycle than any currently produced type of cell.

Cell voltage is greatly affected by temperature. The cell voltage under charge is lower at higher temperatures. At temperatures below 40° F. the cell effectively ceases working. Optimum operating temperature is around 122° F. Figure 10-5 illustrates the effect on temperature and charge rate on the zinc-chlorine cell's voltage. It is possible that the output voltage of a zinc-chloride battery could be controlled by varying the electrolyte's temperature and rate of flow.

Zinc-chloride cell prototypes have capacities between 440 and 2,900 ampere-hours. The capacity of these cells is rated at a C/4 discharge rate. Energy density is over 101 watt-hours per kilogram, which is about three times that of currently produced batteries. The capacity ratings do not include

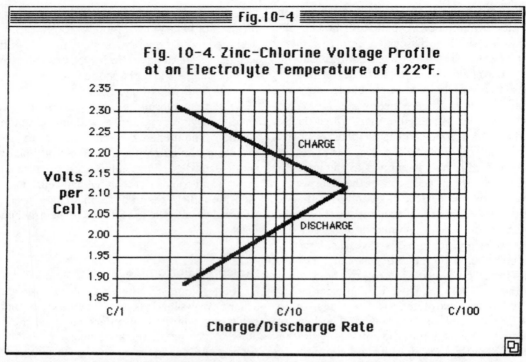

Fig. 10-4. Zinc-Chlorine Voltage Profile at an Electrolyte Temperature of 122°F.

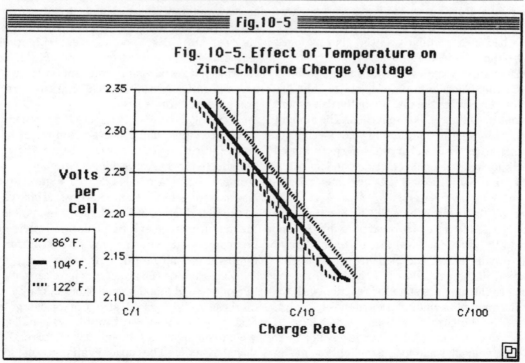

Fig. 10-5. Effect of Temperature on Zinc-Chlorine Charge Voltage

the energy required to drive the necessary pumps and heat exchangers needed to run the battery. The cells when assembled into batteries share common pumps, refrigeration, and electrolyte storage. A control system is used to ride herd on the entire process.

The zinc-chloride cell is about 65 to 75 percent efficient, depending on whether the energy required to operate the pumps and refrigeration is considered. This is comparable with most other types of batteries. The zinc-chlorine battery has an indefinite lifetime. Cleaning and electrolyte replacement will restore the battery to new condition. Cycle lifetimes over 1,000 cycles are currently possible with prototype cells before rebuilding.

There are significant safety problems inherent in the zinc-chloride reaction. The major hazard is the presence of chlorine gas, which is a deadly poison. It is estimated that an electric vehicle battery would contain some 90 pounds of chlorine. Storage of the chlorine in its hydrated form of chlorine hydrate reduces its activity and renders it easier to handle. If the electrolyte storage container were ruptured, as in a vehicular accident, the chlorine gas would be released slowly as the spilled electrolyte warms up. The actual amount of chlorine existing as a gas during actual battery operation will only amount to a few liters. It is possible that the cell may be run at pressures less than atmospheric pressure to reduce the danger of leakage. This type of battery may be difficult to market to motorists who are concerned by the hazards involved. As all types of batteries contain harmful chemicals, it is a matter of choosing the lesser of several evils.

Projected Cost and Availability

Costs are projected to be in the neighborhood of $30 to $40 per kilowatt-hour. This price is very cheap for a battery of such high energy density. This represents a battery with three times the energy density of the lead-acid type at about one-half the cost. Battery cost is projected to be about $ .0075 per mile in vehicular service, with another $ .007 per mile for electricity for recharging. These prices are easily competitive with gasoline as a power source. It is possible that the zinc-chlorine battery may be the first used in commercially manufactured electric vehicles.

The zinc-chlorine cell is projected to be commercially available by the late 1980's. The utility load leveling type of cell is currently being tested and should be available before the vehicular type of cell. There are still many problems to overcome in vehicular service.

SODIUM-SULPHUR

The sodium sulphur cell is one of a family of high temperature batteries. Operating temperatures within the sodium-sulphur cell are between 570° F. (300° C.) and 660° F. (350° C.). Cell design is radically different from low temperature batteries. This cell uses liquid (molten) electrodes and a solid electrolyte. The anode is composed of molten sulphur and sodium polysulphide. The cathode is composed of molten sodium. Both glass and beta-alumina ceramic are used for the solid electrolyte.

Sodium-sulphur prototype cell cases are constructed of metal. The molten sulphur and sodium polysulphide reside in the lower portion of the case. The beta-alumina trough extends into the pool of molten sulphur. This beta-alumina trough is filled with molten sodium. The entire cell is held at a temperature around 600° F. Figure 10-6 shows a cross-sectional drawing of a prototype sodium-sulphur cell. Cells are closely grouped within the battery to retain heat and maximize thermal efficiency. The battery case is heavily insulated to minimize heat loss.

As the cell is cycled, the chemical composition of the sodium polysulphide changes. In the beginning of the discharge cycle, the sodium polysulphide exists as Na_2S_5. As the battery is discharged, the chemical content of the polysulphide gradually changes to Na_2S_3. Overdischarge results in the irreversible formation of Na_2S_2. This battery stores energy by utilizing the different energy levels possible in the sodium-sulphur chemical bond.

Heaters are necessary to initially start the battery and to hold it at operating temperatures dur-

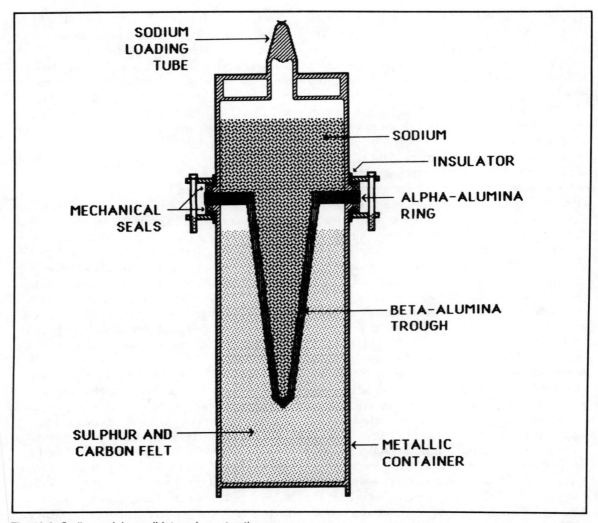

Fig. 10-6. Sodium-sulpher cell internal construction.

ing standby periods. Once the battery is operational, the heat generated by cycling is sufficient to keep the battery at operating temperatures. The heat generated by rapid cycling of the cell is an advantage for the sodium-sulphur cell. Cell prototypes are more efficient if cycled rapidly. The waste heat generated reduces the amount of external thermal energy that must be added to maintain the cell's temperature. If a sodium-sulphur battery is being continually cycled it requires very little, if any, additional heat input.

Projected Performance

The sodium-sulphur battery is being considered for both load leveling and vehicular service. Since the batteries are still in the developmental stage, there is currently little difference between the two types. The sodium-sulphur cell is being optimized for rapid discharge rates. Rates between C/5 and C/3 are being considered as optimum. Rates over C/3 result in cell overheating and necessitate the use of cooling apparatus to maintain acceptable cell temperatures.

Overcharging of the sodium-sulphur cell results in cell destruction. The internal resistance of the cell becomes very high when the cell is fully charged. If additional energy is put into the cell, it is dissipated as heat which raises the cell's internal resistance even further. Overcharging results in thermal runaway and cell destruction. Cell prototypes suffered cracking of the electrolyte trough due to thermal stress.

Cell voltage for the sodium-sulphur reaction is around 2 volts. As with most types of cells, actual cell voltage depends on the rate of discharge and the depth of the cell's discharge. Figure 10-7 shows the relationship between cell voltage and depth of discharge for various discharge rates. This graph is for the sodium-sulphur cell using the beta-alumina electrolyte. The voltage drop experienced as the cell approaches a 50 percent depth of discharge is due to the changing composition of the sodium polysulphide.

Prototype sodium-sulphur cells have a capacity of about 15 ampere-hours. It is expected that the vehicular versions will have capacities in the neighborhood of 1,500 ampere-hours. The utility load leveling models have capacities projected to be around 5,000 ampere-hours. The material and energy requirements of support hardware (heaters) make it cost effective to produce cells with large capacities.

The sodium-sulphur cell is designed to be charged at rates between C/3 and C/10. Rates of charge slower than this are inefficient due to the additional heat required to warm up the battery. Rates higher than C/3 can result in too much heat and thermal runaway. Figure 10-8 represents the relationship between cell voltage and state of charge for various rates of charge. The cell graphed here has a beta-alumina electrolyte.

The sodium sulphur cell is unique in battery technology because of its high efficiency. If efficiency is examined from a purely electronic exchange viewpoint, the cell is nearly 100 percent efficient. In actual service, the efficiency of the sodium-sulphur cell is dependent on the rate of discharge

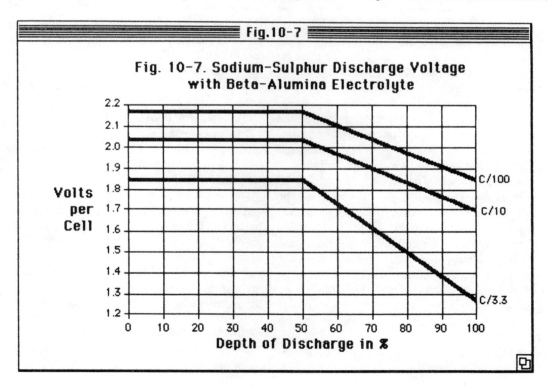

Fig. 10-7. Sodium-Sulphur Discharge Voltage with Beta-Alumina Electrolyte

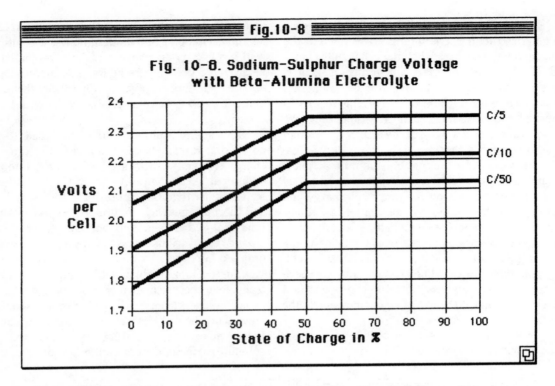

Fig. 10-8. Sodium-Sulphur Charge Voltage with Beta-Alumina Electrolyte

and electrolyte composition. Cells with glass electrolytes being discharged at the C/5 rate have an efficiency of over 90 percent. Figure 10-9 gives the relationship between efficiency and discharge rate for sodium-sulphur cells with glass and beta-alumina electrolytes. The lower efficiency of the beta-alumina electrolyte is due to its higher internal resistance. Heat energy used in starting and maintaining cell temperature are not included in these efficiency estimates. Even if the additional heat energy is considered in the estimate, the sodium-sulphur cell is the most efficient type of battery now under development.

The sodium-sulphur cell is also unique in that it does not discharge itself. Self-discharge rates are so low as to not be measurable. Energy may be stored indefinitely in the cool cell. When the energy is required, the cell is heated and discharging may then begin. The only energy lost is for the heat to warm up the battery.

It is still too early in this battery's development to determine its lifetime. Cell prototypes using beta-alumina electrolytes have lasted over 1,000 cycles. Experimental cells with glass electrolytes are lasting about 200 cycles. It is estimated that sodium-sulphur cells can be developed with lifetimes in excess of 2,000 cycles.

The energy density of the sodium-sulphur has yet to be determined. It is expected that, even including insulation and heaters, this type of battery will have an energy density approaching 10 times that of the lead-acid type.

As with all high temperature batteries, there are safety hazards involved in the sodium-sulphur battery. Molten sodium and molten sulphur are highly reactive elements. In the event of rupture of the electrolyte trough, it is possible that the battery may explode. Research is being conducted using sodium reservoirs and flow restrictors to limit the hazards involved in possible electrolyte failure.

Projected Cost and Availability

The cost of the sodium-sulphur battery has yet to be determined. Cost goals for the program are

$40 per kilowatt-hour. If this estimate can be reached, then the sodium-sulphur cell is definitely viable in utility load leveling applications. It is still too early to predict cost of motive cells using the sodium-sulphur reaction. It is difficult to determine if the average motorist will accept driving about with several hundred pounds of molten sodium and molten sulphur under the hood.

Availability of the sodium-sulphur battery is largely a matter of conjecture. Load leveling types will probably be available first, around the late 1980s. It is impossible to predict when the vehicular type will come to market, certainly not before the mid 1990s.

Alternative energy users are advised to keep abreast of developments in sodium-sulphur battery technology. The high efficiency of this type of battery makes it a candidate for possible alternative energy use, especially in large systems.

LITHIUM-METAL SULPHIDE

The lithium-metal sulphide cell is another type that runs at high temperatures, around 840° F. The internal cell construction is more conventional than the sodium-sulphur cell. The lithium-metal sulphide cell has solid electrodes and a liquid (molten) electrolyte. The electrodes are stacked in plates with separators between them, much in the same manner as currently produced lead-acid cells. The anode of the cell is composed of iron sulphide. The cathode is constructed of an alloy of lithium and aluminum. The electrolyte is a molten mixture of potassium chloride and lithium chloride.

The lithium-metal sulphide cell stores energy by the transfer of sulphur ions from the cathode to the anode. In the discharged cell, the lithium metal on the cathode changes to lithium sulphide, while the anode changes from iron sulphide to iron. In the charged state, the cell's cathode is lithium, and the anode is iron sulphide. This electrochemical reaction will only take place at the elevated temperatures used. The electrolyte is used as an electron transfer medium and does not enter into chemical change with the cell's active materials.

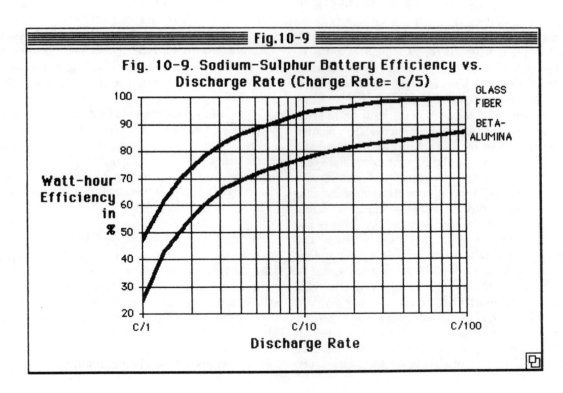

Fig. 10-9. Sodium-Sulphur Battery Efficiency vs. Discharge Rate (Charge Rate= C/5)

Initial experiments with lithium-metal sulphide cells have developed cells with very high energy densities. Due to the highly reactive nature of the active materials, the cells have very short lifetimes. Experiments are being conducted to improve the cell's longevity. Internal corrosion in the cell is a major problem in prototype lithium-metal sulphide cells. Some ideas being tried are separators of boron nitride fabric between the cell's plates and varying the electrolyte's composition.

Projected Performance

The lithium-metal sulphide cell is being optimized for discharge rates around C/3 to C/5. These rates reflect its intended usage as a utility load leveling battery. Prototype lithium-metal sulphide cells are restricted to discharge rates no faster than C/3. Cells designed for more rapid cycling are very short lived and are thus inherently too costly. Charge rates are in the range of C/5 to C/10.

If the lithium-metal sulphide cell is charged and then cooled to room temperature, it will hold its charge indefinitely. The self-discharge rate of the cooled cell is virtually zero. Of course, the cell must be heated to 840° F. (450° C.) in order to be discharged. The cell temperature must be maintained at this high level in order for the battery to be cycled. In commercial power installations waste heat may be used for heating the battery. In areas where waste heat is not available, electric heaters are used to heat the cells. Repetitive cooling and reheating of prototype lithium-metal sulphide cells have reduced the available capacity of the test cells.

Cell voltage of the lithium-metal sulphide cells is around 1.8 volts. Cell voltage is dependent on rate of discharge and state of charge. Cell voltage is also affected by temperature. Increasing the temperature will raise the cell's voltage. Temperatures over 840° F. are inefficient due to greatly increased internal corrosion within the cells. Figure 10-10 illustrates the relationship between cell voltage and state of charge for various discharge rates. The lumps in these curves reflect the various different chemical states of iron sulphide that are produced as the cell is discharged. The charge curves for this cell are the same as the discharge curves, with a voltage factor added for the cell's internal resistance.

Currently developed cell prototypes have capacities in the 50 to 150 ampere-hour range. It is estimated that commercial load leveling models will have capacities around 1,500 ampere-hours. Models of the lithium-metal sulphide cell being developed for vehicular service have capacities around 450 ampere-hours. These capacity figures reflect the energy stored in the battery, not the actual amount of usable energy. Heating requirements are not calculated into the capacity figures.

From an electronic (ampere-hour) viewpoint, the lithium-metal sulphide cell is between 98 and 100 percent efficient. Cell prototypes show an overall electrochemical efficiency of around 70 to 80 percent. These figures do not include any heat added to maintain the battery's temperature. Early cell prototypes exhibit very little loss of efficiency when repeatedly cycled.

Lithium-metal sulphide cell prototypes are lasting for around 400 cycles. It is estimated that cells with cycle lifetimes over 1,000 cycles may be developed. At this point in time, longevity is a major problem with this type of cell. The highly reactive nature of the active materials held at such high temperatures makes rapid deterioration a certainty with this type of cell.

The safety requirements for the lithium-metal sulphide cell are less of a problem than with the sodium-sulphur cell. The reactants are less dangerous to humans. The high temperatures involved in the lithium-metal sulphide cell present a hazard. The cells must be heavily insulated for safety and thermal efficiency.

Projected Cost and Availability

Projected cell costs are between $25 to $35 per kilowatt-hour. If the price of heaters and other peripheral equipment is included, the working system is estimated to cost between $45 to $55 per kilowatt-hour. These prices reflect automated mass production. Current prototypes are many times more expensive.

The availability of the lithium-metal sulphide

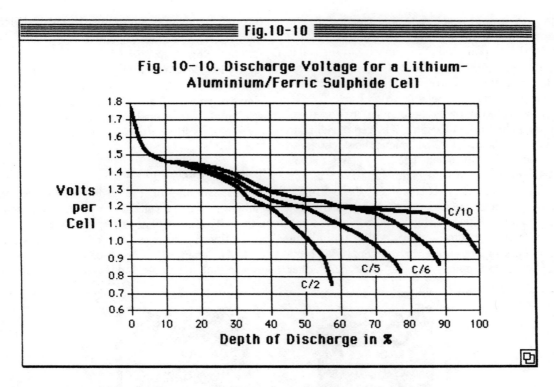

Fig. 10-10. Discharge Voltage for a Lithium-Aluminium/Ferric Sulphide Cell

cell is projected to be around the late 1980s. This estimate is highly optimistic and is based on the solution of the cycle lifetime problem.

REDOX

Redox is an acronym made from the two words reduction and oxidation. These words are the names of the chemical processes involved in most electrochemical cells. The redox cell is an ambient temperature device; it operates at room temperatures. The redox cell uses flowing liquid active materials. The positive liquid is known as the *anolyte*, while the negative fluid is known as the *catholyte*. The fluids are circulated through a reaction cell where they interact. The electrodes of the redox cell are solid inert material. They do not participate in the electrochemical reactions but are merely sites for the reactions to take place. The electrodes are separated by an ion-selective membrane. The anolyte and catholyte are pumped through the reaction cell for charging and discharging. The anolyte and catholyte react on the inert electrodes and are stored in external and separate tanks.

Prototypes of the redox cell use chromium dichloride ($CrCl_2$) dissolved in water as an anolyte. The catholyte is composed of ferric chloride ($FeCl_3$) dissolved in water. The ion selective membrane allows the flow of the negatively charged chlorine ion, while inhibiting the flow of positively charged chromium and iron ions. During the discharge cycle of the redox cell, the catholyte is reduced while the anolyte is oxidized. Chloride ions migrate through the ion selective membrane. During the charge portion of the cycle, the chemical process is reversed. The catholyte is oxidized while the anolyte is reduced. During the discharge portion of the cycle, the negative chlorine ions migrate to the inert cathode. During the charge portion of the cycle, the negatively charged chlorine ions migrate through the membrane to the anode.

Figure 10-11 illustrates the workings of a two-tank electrically rechargeable redox flow cell. There are two separate tanks for anolyte and catholyte storage, each with its own pump. The energy output limit of the cell is determined by the size of the reaction cell. The capacity of the cell is determined by the size of the storage tanks. This feature is unique in batteries. The redox cell bears a close physical resemblance to fuel cells.

Most batteries evolve amounts of hydrogen gas at the anode during charging. The redox cell is no exception. The redox cell is unique, however, in its usage of the hydrogen gas. The hydrogen gas is collected and directed to a chamber known as the *rebalance cell*. In the rebalance cell, the hydrogen reacts with the same iron solution used in the rest of the system. This additional reaction adds some 0.7 volts to the cells' output, while minimizing hydrogen loss from the entire system. This recycling of the hydrogen gives the redox cell higher efficiency, greater longevity, and reduced maintenance.

The redox cell is being developed primarily as a utility load leveling battery. Its energy density is too low for it to be considered as a prospect for vehicular power. The redox cell is still very experimental, and as such, many parameters of its operation are still not definitely settled.

Projected Performance

Rest voltage of the redox cell is in the neighborhood of 1.2 volts. The rest voltage depends on the cell's state of charge. Figure 10-12 illustrates the relationship between the cell's voltage and its depth of discharge. The redox cell is like other batteries in that its voltage under discharge is propor-

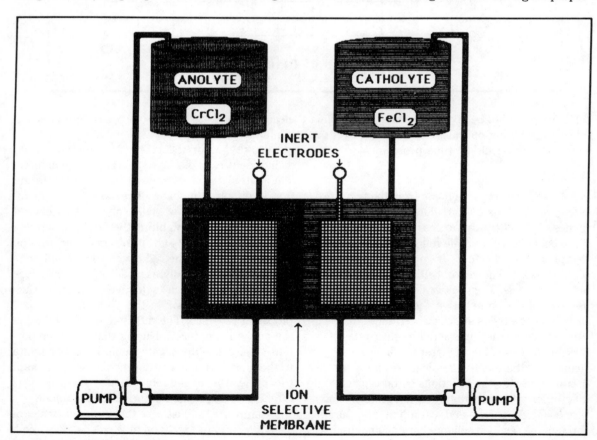

Fig. 10-11. REDOX cell schematic.

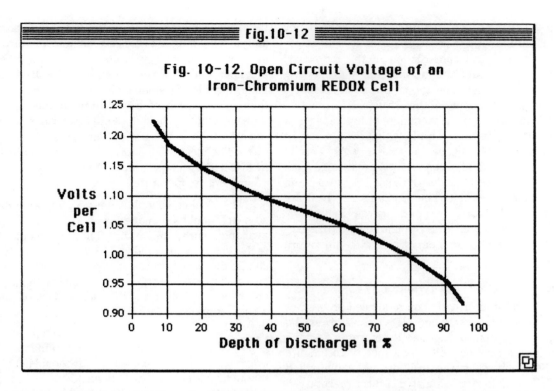

Fig. 10-12. Open Circuit Voltage of an Iron-Chromium REDOX Cell

tional to the discharge rate. The cell voltage must be higher than the rest voltage in order to charge the cell. This is a function of the cell's internal resistance. In the redox cell, the internal resistance is primarily a function of the membrane resistivity. Much of the development being done on the redox cell is focused on developing a low resistance ion selective membrane.

The active materials of the redox cell are carried in the storage tanks. All cells within the battery use the same anolyte and catholyte tanks. This feature makes it simple to add or delete cells from the battery without introducing imbalances within the cells. The addition of cells to an already running battery makes it possible to regulate the voltage output of the battery regardless of its state of charge. This is a highly desirable feature in utility load leveling batteries.

The capacity of the redox cell is determined by the amount of the active materials present. This is a function of the size of the anolyte and catholyte storage tanks. Current prototype cells have a reaction cell output of 100 watts, with tanks which give it a capacity of 400 watt-hours. The cost requirements of auxiliary equipment such as pumps and tanks make it inefficient to use cells with capacities smaller than 150 kilowatt hours. The redox cell battery must have considerable size in order to be cost effective. The redox cell is not damaged by complete discharging or by overcharging.

The redox cell is capable of sustained high charge rates of over C/3. This is due to the hydrogen absorption in the rebalance cell. The redox cell's ability to be quickly and efficiently recharged is a major advantage in utility service. When the cell is rapidly recharged, the pumps are simply run faster to increase the flow of the active materials through the reaction cell.

The redox cell is estimated to be about 80 percent efficient. This figure reflects one of the major problems still to be overcome in the redox cell—membrane leakage. The current prototypes of redox cell membranes allow slight crossover of iron and chromium ions, resulting in lower efficiency and cell life.

The physical construction of the redox cell results in an indefinite lifetime. Ion migration between the anolyte and the catholyte necessitates periodic replacement of these fluids. The reaction cell and other auxiliary equipment are capable of unlimited operation. It is likely that the membrane will have to be replaced about every 20 years to minimize iron and chromium ion cross migration.

The redox cell is one of the safest types being currently developed. The reactants involved are neither highly corrosive nor dangerous to humans. Since the redox cell is an ambient temperature battery, there are no problems with high temperature materials.

Projected Cost and Availability

The cost information on redox cell prototypes is divided into two categories—output power and capacity. Output power cost is determined by the size of the reaction cell. The capacity cost figure is determined by the size of the anolyte and catholyte tanks. It is estimated that the reaction cell will cost about $160 per kilowatt. The capacity estimate is $20 per kilowatt-hour. If the battery is assumed to have an output of ten kilowatts and a capacity of 400 kilowatt-hours, then the total cost would be about $24 per kilowatt-hour. This price is very low, which explains the interest in redox cell.

The redox cell is projected to be available by the early 1990s. This estimate assumes solution of the ion-selective membrane's relatively high resistance and the ion cross migration problem.

THE IDEAL BATTERY

The ideal battery should have a number of different characteristics. It should have, first of all, a high cell voltage. The electrochemical nature of battery stored energy limits cell voltage to the neighborhood of two or three volts. The discharge voltage curve of the ideal cell should be very flat. It should maintain a fairly constant voltage until it is nearly empty. The output characteristics should remain stable over a wide variety of discharge rates. The ideal cell's output voltage should be little affected by temperature. The cell must operate at ambient temperatures.

The ideal cell should have a high energy density. It should be made from light, commonly available, cheap materials. The cell should be rechargeable and capable of many thousands of charge/discharge cycles. It should be able to store energy for extended periods of time without appreciable self-discharge. The ideal cell should also offer a long calendar lifetime of constantly efficient service.

In addition to all the above requirements, the cell must be safe to use and store. It should not contain toxic materials. The cell must be so effectively contained that leakage is impossible. It is highly probable that the ideal electrochemical cell will contain lithium. Lithium is a component in all the electrochemical couples which offer characteristics close to the ideal cell. Lithium is both highly electrically reactive and very light. It is also relatively cheap and available.

It is apparent that the ideal cell is a close relative to the elusive "free lunch." Both seem highly desirable and equally improbable. Modern technology is working hard to give us better batteries. How close these new cells come to the ideal cell is a matter for future discussion.

Appendix A

Formulas and Conversion Factors

OHM'S LAW

$$E = IR$$

E = Electromotive Force expressed in Volts
I = Current expressed in Amperes
R = Resistance expressed in Ohms (Ω)

POWER

$$P = IE$$

P = Power expressed in Watts
I = Current expressed in Amperes
E = Electromotive Force expressed in Volts

$$P = I^2R$$

P = Power expressed in Watts
I = Current expressed in Amperes
R = Resistance expressed in Ohms (Ω)

RESISTANCE IN SERIES

$$R_t = R_1 + R_2 + R_2 + \ldots R_n$$

R_t = Total Resistance expressed in Ohms (Ω)
R_1 to R_n = Individual Resistances expressed in Ohms (Ω)

RESISTANCES IN PARALLEL

$$\frac{1}{R_t} = \frac{1}{R_1} + \frac{1}{R_2} + \frac{1}{R_3} + \cdots \frac{1}{R_n}$$

R_t = Total Resistance expressed in Ohms (Ω)
R_1 to R_n = Individual Resistances expressed in Ohms (Ω)

CHARGE/DISCHARGE RATE

$$I = \frac{C}{T}$$

I = Rate of charge or discharge expressed in amperes
C = Battery's rated capacity expressed in ampere-hours
T = Cycle time period expressed in hours

SPECIFYING WIRE SIZE

$$R = \frac{E}{IL} (1000)$$

R = Maximum Resistance expressed in Ω per 1000 feet
E = Maximum allowable voltage drop in the wiring expressed in volts
I = Amount of current flow through the wire expressed in amperes
L = The length of the wire in the complete circuit expressed in feet

SIZING BATTERY CAPACITY

$$C = \frac{(Pd)(D)(1.25)}{Vb}$$

C = Capacity of the battery pack expressed in ampere-hours
Pd = Estimated power consumption expressed in watt-hours per day
D = Time between battery charges expressed in days
Vb = Voltage of the battery pack expressed in volts

CONVERSION OF HORSEPOWER TO WATTS

$$Pw = Ph (735.5)$$

Ph = Power expressed in Horsepower
Pw = Power expressed in Watts

Appendix B
Forms

POWER COST ESTIMATE

DC POWER CONSUMPTION in W-hrs./day	
AC POWER CONSUMPTION in W-hrs./day	
TOTAL POWER CONSUMPTION in W-hrs./day	

NUMBER OF DAYS OF BATTERY STORAGE	

BATTERY CAPACITY in AMPERE-HOURS		
BATTERY PACK VOLTAGE		
ESTIMATED BATTERY COST		

ESTIMATED INVERTER WATTAGE		
ESTIMATED INVERTER COST		

ESTIMATED COST: INVERTER & BATTERIES	

POWER SOURCE		
POWER SOURCE WATTAGE		
ESTIMATED POWER SOURCE COST		
ESTIMATED OPERATING COST PER YEAR		
ESTIMATED SOURCE LONGEVITY IN YEARS		

ESTIMATED INITIAL HARDWARE COST	

ESTIMATED COST PER YEAR OVER A 10 YEAR PERIOD	
ESTIMATED ENERGY COST IN $ PER KILOWATT-HOUR	

DC POWER ESTIMATION

DEVICE	DEVICE WATTAGE	ON TIME PER DAY	WATT-HRS. PER DAY
	DC POWER CONSUMED DAILY - WATT-HOURS ▶		

AC POWER ESTIMATION

DEVICE	STARTING SURGE	DEVICE WATTAGE	ON TIME PER DAY	WATT-HRS. PER DAY
TOTAL SURGE WATTAGE ➤		**AC POWER CONSUMED DAILY - WATT-HOURS** ➤		
		PLUS _____ % FOR INVERTER INEFFICIENCY		
		AC POWER CONSUMED DAILY WATT HOURS ➤		

Glossary

active material—The materials which chemically react within the cell to release free electrons are known as active materials. In most cases, one active material is a metal or metallic compound which is oxidized. The other active material, often a metallic oxide, is reduced.

ampere—The ampere is the standard unit used to measure electrical current. Physically, the ampere is a measure of the number of electrons passing a given point per unit time.

ampere-hour—The ampere-hour is the unit of measurement of the electrical capacity of a cell or battery. Physically, it represents the number of electrons available from the cell or battery.

anode—The anode is the electrode within the cell which undergoes the chemical process of oxidation. Electrically, the anode is the cell's positive terminal.

antimony—Antimony is a metallic chemical element with the atomic number of 51. Antimony is alloyed with lead to physically strengthen the plates of lead-acid cells.

battery—A battery is a group of interconnected electrochemical cells. Single cells are considered to be a battery if they are used alone.

capacity—Capacity is the amount of electrical energy a cell or battery contains. The ampere-hour is the unit of this capacity.

cathode—The cathode is the electrode within the cell which undergoes the chemical process of reduction. Electrically, the cathode is the negative terminal of the cell.

cell—The cell is the basic unit used to store energy in the battery. The cell contains an anode, a cathode, and the electrolyte.

cutoff voltage—The voltage level at which a cell is considered to be empty, and the discharge process is terminated.

cycle—A cycle is one complete charge/discharge sequence of the battery.

cycle life—Cycle life is the number of cycles a cell or battery will undergo before being considered "worn out". This point is usually defined as

181

when the battery's capacity has reached only 80 percent of its initial rated capacity.

deep cycle—A battery or cell is said to be "deep cycled" if 80 percent or more of its energy is withdrawn before recharging.

dendrites—Dendrites are microscopic whiskers of metal which form in nickel-cadmium cells. These metallic whiskers may cause internal shorting within the cell, rendering it useless.

depth of discharge—The amount of energy withdrawn from a battery or cell expressed as a percentage of its rated capacity.

electrochemical couple—An electrochemical couple is two chemical compounds or elements which react together to release free electrons.

electrolyte—The electrolyte is the medium of ion transport within the cell. The electrolyte provides a path for electron transfer between the anode and cathode of the cell. Electrolytes are usually liquids or pastes, which are either acidic or basic.

end of charge voltage—The voltage level at which a cell or battery is considered, while under charge, to be full.

energy density—Energy density is a ratio of a battery or cell's capacity to either its volume or weight. Volumetric energy density is expressed in watt-hours per cubic inch. Weight energy density is expressed in watt-hours per pound.

energy transfer rate—The energy transfer rate is a measure of the rate at which energy is either being added or withdrawn from a battery or cell. This energy transfer is measured in amperes.

equalizing charge—The equalizing charge is a controlled overcharge of an already full battery to restore all the individual cells within the battery to the same state of charge.

float service—A battery is in float service when it is continually charged at a very slow rate, and only occasionally discharged.

gassing—Gassing is the evolution of hydrogen and oxygen gases at the cell's electrodes. These gases result from the hydrolysis of water in the electrolyte during the charging process.

grid—The grid within a cell is an electrically conductive structure which holds the cell's active materials. The grid may or may not participate in the chemical reactions of the cell.

hydrometer—The hydrometer is an instrument for measuring the density of liquids in relation to the density of water. The hydrometer is used to indicate the state of charge in lead-acid cells by measuring the specific gravity of the electrolyte.

ion—An ion is an electrically charged particle or molecule.

local action—Local action is the process of self-discharge that is present in all forms of electrochemical cells.

primary cell—A primary cell is an electrochemical cell which cannot be recharged. The chemical process within the primary cell is only one way—discharge. When a primary cell is discharged it is discarded.

rate of charge—The amount of energy per unit time that is being added to the battery. Rate of charge is commonly expressed as a ratio of the battery or cell's rated capacity to charge duration in hours.

resistance—Resistance is the property of materials to impede a flow of electrons through themselves. All materials have some resistance. Those of low resistance are known as conductors, while those of high resistance are known as insulators. The unit used to measure resistance is the Ohm (Ω).

rest voltage—The voltage of a cell or battery that is neither being charged or discharged.

secondary cell—Secondary cells are electrochemical cells which are rechargeable. The chemical reaction within the secondary cell is reversible, allowing the cell to be recharged many times.

self-discharge—Self-discharge is the tendency of all electrochemical cells to lose energy. Self-discharge represents energy lost to internal chemical reactions within the cell. This energy is not and cannot be used from the battery or cell's output terminals.

specific gravity—Specific gravity is the ratio of a liquid's density to the density of water.

state of charge—State of charge is a ratio, expressed in percent, of the energy remaining in a battery in relation to its capacity when full.

sulphation—Sulphation is the formation of lead sulphate crystals on the plates of lead-acid cells. Some of these crystals are bonded in a covalent manner and are impossible to re-ionize during recharging. As such, they represent a loss in capacity to the cell.

volt—The volt is the unit used in the measurement of the electromotive force. A standard electrical definition of the volt is: an electromotive force of 1 volt is necessary to move a current of 1 ampere through a 1 Ω resistor.

watt—The watt is the unit used to measure power. In electrical terms, it is a volt-ampere.

Index

A
active materials, 4
adiabatic reaction, 40
alkaline dry cell, 72
alkaline-manganese cells, 77
alkaline-manganese cells, chemical and physical construction, 78
alkaline-manganese cells, cost, 82
alkaline-manganese cells, energy specifications, 79
alkaline-manganese cells, shelf life, 80
alkaline-manganese cells, temperature effects, 81
alkaline-manganese cells, types of service, 81
alternating current, 2
amperage, 2
ampere, 2
anode, 12
anolyte, 171
appliance power consumption, 135
appliances, 146
arsine, 36
automobile alternators, 101
automotive alternator control systems, 103
automotive starting batteries, 14
average power consumption, 136

B
batteries in solar systems, equalizing, 101
batteries suitable for inverter use, 132
batteries, low antimony deep cycle, 15
batteries, storing and transferring energy, 4
batteries, true deep cycle high antimony, 15
batteries, using effectively, 107-123
battery instruments, 121
battery maintenance during recharging, 30
battery pack, internally cross-wiring, 115
battery pack, sizing, 142
battery technologies, developing, 159-174
battery types, 7
battery, definition of, 1-13
battery, efficiency of, 6
battery, location in system, 120

C
cadmium hydroxide, 59
capacity, 2
capacity, as a function of age, 25
capacity, as a function of temperature, 21
cathode, 12
catholyte, 171
cell polarity nomenclature, 12
cell, 4
cells, 7
cells, assembling into batteries, 7
cells, in parallel for capacity increase, 10
cells, in series for voltage increase, 9
charge reactions, 33
chargers, 120 volt ac powered, 9
chemical composition, 33
circuit breakers, 114
conductors, 2
connections, 112
copper wire table, 109
covalent bonds, 34
current, 2

D
Davies, Humphrey, 70
dendrites, 57
dendritic growth, 57, 61
depth of discharge, 32
digital multimeter (DMM), 121
direct current, 2
discharge cutoff voltage, 45
discharge rate, 31
discharge reactions, 33
discharging, 5
dry cells, 72

E
Edison cell characteristics, 64
Edison cell, chemical composition, 68
Edison cell, discharging, 67
Edison cell, life expectancy, 64
Edison cell, technical data for, 68
Edison cells, 63-69
Edison cells, applications for, 64
Edison cells, charging, 65
Edison cells, cost, 64
Edison cells, physical construction, 63
Edison cells, temperature, 65
Edison cells, voltage, 65
Edison, Thomas A., 63
efficiency, 2
efficiency, as a function of discharge rate, 22
electric motors, 147
electrochemical couples, 4
electrolyte stratification, 31
electrolyte, 34
electromotive force (EMF), 1

electronics, 148
endothermic heat, 37
energy consumption, estimating, 134
energy estimate for a large system, 138
energy estimate for a small system, 136
energy management techniques, 144
energy management, 134
energy management, using to reduce system cost, 149
energy storage in chemical reactions, 5
energy transfer, rate of, 7
equalizing charge, 30
exothermic heat, 37

F
Faraday, Michael, 71
filtration, 122
flashlight battery, 73
float charging, 30
forms, 178-180
formulas and conversion factors, 175-176
fuses, 112, 114

G
gassing, 36
gel cells, 16

H
hydrometer, 4

I
insulators, 2
internal resistance, 5
inverter for a large system, 141
inverter for a small system, 139
inverter input voltage, 141
inverter location, 133
inverter sizing, 131
inverter to battery interconnect, 132
inverter, how it works, 124
inverter, sizing for average power consumption, 139
inverter, sizing for peak power consumption, 139
inverter, sizing, 139
inverters with battery chargers, 130
inverters, 124-133
inverters, different types of, 125
inverters, modified sine wave, 127
inverters, pure sine wave, 130
inverters, square wave, 126

L
lawnmower engines, 101

184

lead-acid batteries, 14-40, 144
lead-acid batteries, discharging, 31
lead-acid battery chargers, 96
lead-acid battery, characteristics of, 17
lead-acid battery, charging, 26
lead-acid battery, technical data, 33
lead-acid couple, 7
lead-acid deep cycle battery manufacturers, 40
Leclanche, Georges, 71
lighting, 146
lithium cell, 72
lithium cells, 90
lithium cells, chemical and physical construction, 90
lithium cells, temperature effects, 93
lithium cells, types of service, 94
lithium-metal sulphide, 169
local action, as a function of temperature, 24
longevity, as a function of depth of cycle, 22

M

mechanical connections, 113
mercury cells, 82
mercury cells, chemical and physical construction, 82
mercury cells, cost, 85
mercury cells, energy specifications, 83
mercury cells, shelf life, 84
mercury cells, temperature effects, 84
mercury cells, types of service, 85
methods and machines to charge batteries, 96-106
motorized charger for 12-volt systems, 101

N

ni-cad batteries, sizing, 58
ni-cad charger, battery powered, 105
ni-cads, memory effect, 47
ni-cads, rejuvenation of tired, 56
nickel hydroxide, 59
nickel oxide hydroxide, 59
nickel-cadmium batteries, 41-62
nickel-cadmium batteries, 7
nickel-cadmium battery characteristics, 45
nickel-cadmium battery chargers, 96
nickel-cadmium battery manufacturers, 62
nickel-cadmium cell, charging, 50
nickel-cadmium cell, discharging, 58
nickel-cadmium cell, contamination and water loss, 60
nickel-cadmium cells, technical data for, 59
nickel-cadmium chargers, commercial, 106

nickel-cadmium chargers, small, 104
nickel-zinc cell, 160

O

Ohm's Law and wiring resistance, 108
Ohm's Law, 2
outlets, 112
outlets, 114

P

parallel wiring, 7
peak power consumption, 135
plugs, 114
pocket plate ni-cad applications, 44
pocket plate ni-cads (vented and sealed), 43
power, 2
primary cell manufacturers, 94
primary cells, 70-95
pumps, 148
purity, 6

R

rate of charge or discharge, 4
rebalance cell, 172
redox reaction, 60
redox, 171
refrigeration, 147
renewal solution, 69
resistance, 2
rest voltage, as a function of state of charge, 48
rotor, 102
Rubin, Samuel, 72

S

safety requirements, 120
self-discharge or local action, 32
series and parallel interconnection, 10
series wiring, 7
silver oxide cell, cost, 88
silver oxide cell, types of service, 88
silver oxide cells, 86
silver oxide cells, chemical and physical construction, 86
silver oxide cells, energy specifications, 87
sintered plate ni-cad applications, 42
sintered plate ni-cads (vented and sealed), 41
sizing, 6
sodium-sulphur cell, 165
solar array sizing, 98
solar cell regulators, 100
solar cells, 98
soldered connections, 113
specific gravity, 2
specific gravity, as a function of state of charge, 21
state of charge, 2

state of discharge, 2
stator, 102
stibine, 36
sulphation, 34
switches, 112
switches, 114
switching, 125
system cost estimates, 149

T

temperature, 6
thermal runaway, 52
transformer, 125
Trojan L-16, 11

U

uninterruptible power supplies (UPS), 130

V

volt, 1
Volta, Allesandro, 70
voltage, 1
voltage, as a function of state of charge, 19
voltmeter, 121
voltmeter, expanded scale analog, 121

W

whiskers, 61
wind and water power sources, 103
wire size, techniques for specifying, 111
wire, using for current measurement, 117
wiring loss, calculating, 109
wiring techniques, low voltage, 112
wiring, 108

Z

zapping, 57
zinc-air cells, 89
zinc-air cells, chemical and physical construction, 89
zinc-air cells, cost, 90
zinc-air cells, energy specifications, 89
zinc-air cells, shelf life, 90
zinc-air cells, types of service, 90
zinc-carbon cell, cost, 77
zinc-carbon cell, effects of temperature, 76
zinc-carbon cell, energy specifications, 75
zinc-carbon cell, shelf life, 75
zinc-carbon cell, types of service, 76
zinc-carbon cells, 73
zinc-carbon cells, chemical and physical construction, 73
zinc-carbon cells, energy specifications, 75
zinc-carbon cells, shelf life, 75
zinc-chlorine cell, 162

OTHER POPULAR TAB BOOKS OF INTEREST

Transducer Fundamentals, with Projects (No. 1693—$14.50 paper; $19.95 hard)

Second Book of Easy-to-Build Electronic Projects (No. 1679—$13.50 paper; $17.95 hard)

Practical Microwave Oven Repair (No. 1667—$13.50 paper; $19.95 hard)

CMOS/TTL—A User's Guide with Projects (No. 1650—$13.50 paper; $19.95 hard)

Satellite Communications (No. 1632—$11.50 paper; $16.95 hard)

Build Your Own Laser, Phaser, Ion Ray Gun and Other Working Space-Age Projects (No. 1604—$15.50 paper; $24.95 hard)

Principles and Practice of Digital ICs and LEDs (No. 1577—$13.50 paper; $19.95 hard)

Understanding Electronics—2nd Edition (No. 1553—$9.95 paper; $15.95 hard)

Electronic Databook—3rd Edition (No. 1538—$17.50 paper; $24.95 hard)

Beginner's Guide to Reading Schematics (No. 1536—$9.25 paper; $14.95 hard)

Concepts of Digital Electronics (No. 1531—$11.50 paper; $17.95 hard)

Beginner's Guide to Electricity and Electrical Phenomena (No. 1507—$10.25 paper; $15.95 hard)

750 Practical Electronic Circuits (No. 1499—$14.95 paper; $21.95 hard)

Exploring Electricity and Electronics with Projects (No. 1497—$9.95 paper; $15.95 hard)

Video Electronics Technology (No. 1474—$11.50 paper; $16.95 hard)

Towers' International Transistor Selector—3rd Edition (No. 1416—$19.95 vinyl)

The Illustrated Dictionary of Electronics—2nd Edition (No. 1366—$18.95 paper; $26.95 hard)

49 Easy-To-Build Electronic Projects (No. 1337—$6.25 paper; $10.95 hard)

The Master Handbook of Telephones (No. 1316—$12.50 paper; $16.95 hard)

Giant Handbook of 222 Weekend Electronics Projects (No. 1265—$14.95 paper)

Introducing Cellular Communications: The New Mobile Telephone System (No. 1682—$9.95 paper; $14.95 hard)

The Fiberoptics and Laser Handbook (No. 1671—$15.50 paper; $21.95 hard)

Power Supplies, Switching Regulators, Inverters and Converters (No. 1665—$15.50 paper; $21.95 hard)

Using Integrated Circuit Logic Devices (No. 1645—$15.50 paper; $21.95 hard)

Basic Transistor Course—2nd Edition (No. 1605—$13.50 paper; $19.95 hard)

The GIANT Book of Easy-to-Build Electronic Projects (No. 1599—$13.95 paper; $21.95 hard)

Music Synthesizers: A Manual of Design and Construction (No. 1565—$12.50 paper; $16.95 hard)

How to Design Circuits Using Semiconductors (No. 1543—$11.50 paper; $17.95 hard)

All About Telephones—2nd Edition (No. 1537—$11.50 paper; $16.95 hard)

The Complete Book of Oscilloscopes (No. 1532—$11.50 paper; $17.95 hard)

All About Home Satellite Television (No. 1519—$13.50 paper; $19.95 hard)

Maintaining and Repairing Videocassette Recorders (No. 1503—$15.50 paper; $21.95 hard)

The Build-It Book of Electronic Projects (No. 1498—$10.25 paper; $18.95 hard)

Video Cassette Recorders: Buying, Using and Maintaining (No. 1490—$8.25 paper; $14.95 hard)

The Beginner's Book of Electronic Music (No. 1438—$12.95 paper; $18.95 hard)

Build a Personal Earth Station for Worldwide Satellite TV Reception (No. 1409—$10.25 paper; $15.95 hard)

Basic Electronics Theory—with projects and experiments (No. 1338—$15.50 paper; $19.95 hard)

Electric Motor Test & Repair—3rd Edition (No. 1321—$7.25 paper; $13.95 hard)

The GIANT Handbook of Electronic Circuits (No. 1300—$19.95 paper)

Digital Electronics Troubleshooting (No. 1250—$12.50 paper)

TAB TAB BOOKS Inc.

Blue Ridge Summit, Pa. 17214

Send for FREE TAB Catalog describing over 750 current titles in print.